THE PSYCHOLOGICAL APPEAL OF GARDENS

Clive R. Hollin

 Routledge
Taylor & Francis Group

LONDON AND NEW YORK

Designed cover image: © Halfpoint Images/Getty Images

First published 2024
by Routledge
4 Park Square, Milton Park, Abingdon, Oxon OX14 4RN

and by Routledge
605 Third Avenue, New York, NY 10158

Routledge is an imprint of the Taylor & Francis Group, an informa business

© 2024 Clive R. Hollin

British Library Cataloguing-in-Publication Data
A catalogue record for this book is available from the British Library

Library of Congress Cataloging-in-Publication Data
Names: Hollin, Clive R., author.
Title: The psychological appeal of gardens / Clive R. Hollin.
Description: Milton Park, Abingdon, Oxon ; New York, NY : Routledge, 2024. |
Includes bibliographical references and index. |
Identifiers: LCCN 2023022132 (print) | LCCN 2023022133 (ebook) |
ISBN 9781032267265 (hardback) | ISBN 9781032265414 (paperback) |
ISBN 9781003289661 (ebook)
Subjects: LCSH: Gardening--Psychological aspects. | Gardening--Health aspects. |
Gardens--Psychological aspects.
Classification: LCC SB454 .H67 2024 (print) | LCC SB454 (ebook) |
DDC 635.01/9--dc23/eng/20230726
LC record available at https://lccn.loc.gov/2023022132
LC ebook record available at https://lccn.loc.gov/2023022133

ISBN: 978-1-032-26726-5 (hbk)
ISBN: 978-1-032-26541-4 (pbk)
ISBN: 978-1-003-28966-1 (ebk)

DOI: 10.4324/9781003289661

Typeset in Sabon
by MPS Limited, Dehradun

*For Felicity, my wife and personal
gardening consultant*

CONTENTS

PREFACE

The weight of my published work has been in the field of psychology applied to understanding and reducing anti-social and criminal behaviour. After retiring from academia and its stifling government-driven demands for a narrow research output by which to allocate resources, I determined to take a foray into new territory. Accordingly, I set about a new project, which became *An Introduction to Human-Animal Relationships: A Psychological Perspective*, published by Routledge in 2021. This present book is the second in my venture into new terrain and its writing confirmed my experiences with its predecessor. First, in a new field there is an awful lot of background material to become familiar with: sometimes this is tedious and seemingly never ending, at other times it is absorbing and thought-provoking. Second, the increase in background reading necessarily slows down the pace of writing so that the book stays with you for a longer period. In my area of expertise, I would expect to complete writing a book in about a year, now it is getting on for double that time. The need to check and double-check the text also adds to the longer writing time with many more edits than might generally be the case. I've never been a writer who seeks the opinions of other people on a work in progress: I see this as a double-edged sword, you may get some useful comments but, and here I speak from experience, it is possible to be blown off-course by injudicious remarks intended to be helpful but which actually serve to muddy the waters. I much prefer to trust to my own experience and leave the rest to the copyeditor.

As may be apparent I am a keen gardener, following in my paternal grandfather's steps, with a relatively large suburban garden that absorbs a great deal of my time. One of the antecedents to this book was a conversation in which someone said, "I just don't get it with gardening. Why on earth would anyone want to do that?" This remark stayed with me and some initial browsing revealed that there was not a huge amount of psychological research from which to formulate an answer. (The exception to this point lies in the notion of gardening as therapy, which is covered here in some detail.) Incidentally, as Hondagneu-Sotelo (2010)[1] notes, sociology has also largely neglected the potential of the garden to inform us about people and their activities.

Well, over time a few cursory notes grew into a book proposal until, well, here we are.

Although I do not share my writing before it is sent off to the publisher, I do share my thoughts from time to time with my immediate family. My wife's parents were keen and highly capable gardeners and she has an implicit knowledge on which I draw from time to time. My son and my daughter are nascent gardeners: I recognise the obstacles they face in young children and developing careers as I quietly try to reinforce their horticultural efforts. They have all been helpful in different ways and I am grateful to them for that and a great deal more.

Note

1 Hondagneu-Sotelo, P. (2010). Cultivating questions for a sociology of gardens. *Journal of Contemporary Ethnography, 39*(5), 498–516.

INTRODUCTION

The website *Love The Garden* suggests that in the United Kingdom about 27 million people, from a population of 64 million, engage in gardening. It has been estimated that in 2017 gardeners spent about £7.5 billion on garden products, with a further £2.4 billion paid out for the services of gardeners and landscapers. These figures indicate that in an average year each "gardening household" in the United Kingdom spends approximately £150 on their garden, although many gardeners will spend considerably more. Overall, gardening engages over one-third of the country's population and is a significant part of the economy. Given the ubiquitous presence in our everyday lives of gardens and gardening, what is their appeal? Why do we spend significant amounts of our money on plants and gardening paraphernalia? Why do we spend hours of our time tending our annuals, bulbs, and shrubs? Why do we travel to see gardens in both our own and other countries? We can apply psychological knowledge to help formulate answers to these questions with a focus principally, but not exclusively, on the private garden.

DOI: 10.4324/9781003289661-1

1

HOW DID YOUR GARDEN GROW?

The nature of the relationship between *homo sapiens* and their environment has a long history. Some 300,000 years ago our ancestors survived by foraging for edible plants, hunting animals and gathering shellfish. These early humans learned to make their lives easier by protecting useful trees, such as hazel, introducing desirable new species and uprooting unwanted plants. This prehistoric *forest gardening* took place on riverbanks in rainforests and in the foothills of monsoon regions. The exact point when humans moved from cultivating the land to growing food crops to tending a garden where the plants give pleasure is lost in the mists of time. Nonetheless, we know that the history of gardening encompasses the biblical Garden of Eden, the Hanging Gardens of Babylon, pre-Islamic Paradise Gardens, the physic garden and the extensive plant collections to be found in botanical gardens (Spencer & Cross, 2017).

The next step after forest gardening, around 10,000 BCE, was to enclose land, probably to keep out wild animals. This practise began in West Asia and moved westward into Greece, Germany, France, Britain and Spain. It was within these enclosures that garden design and construction gradually emerged. In Anglo-Saxon Britain (circa 450–1150 CE) Old English was the common language: the Old English word *geard*, meaning an enclosed place, is the root of the modern-day *garden*. The changes in human behaviour that eventually led to farming and herding animals had the effect of modifying the natural landscape and, given that farming is not compatible with a nomadic lifestyle, people settled in hamlets and villages, some of which grew into towns and cities.

It may be that the first gardens, in the sense that we would recognise a garden today, were found in ancient Egypt. There are Egyptian tomb paintings dating to the sixteenth century BCE that show lotus ponds encircled by rows of

DOI: 10.4324/9781003289661-2

acacias and palms, indicating the existence of decorative horticulture. The Egyptians had both temple gardens, as at Karnak, and domestic gardens. In the Egyptian heat people sought refuge from the sun, and they created walled gardens with shade-giving trees – date palms, fig trees, olive trees, nut trees, pomegranate trees, sycamores and willows – vineyards for wine production, and flowers, including cornflowers, daisies, irises, poppies and roses. A well-stocked rectangular pond completed the design.

As the centuries rolled by, the great civilisations created their own idiosyncratic gardens. In the sixth century BCE, the Babylonian King Nebuchadnezzar II constructed the Hanging Gardens as a gift to his wife Amytis, a monument celebrated as one of the Seven Wonders of the Ancient World. The reign of Darius the Great (550–486 BCE), King of Persia and ruler of the Achaemenid dynasty, saw the emergence of the *paradise garden*. This style of formal enclosed garden, familiar in the Islamic world, is a rectangular plot traditionally divided into four quarters – reflecting the four compass points, the four seasons, and the earth's four elements – with a pond placed in the centre (Figure 1.1). The scent of fruit trees and quiet areas for peaceful reflection make these gardens a manifestation of heaven on earth, a vision of paradise. In a different religious tradition, the association of gardens with tranquillity is found in the biblical story of Adam and Eve, who lived perfect lives, for a while at least, in The Garden of Delight.

FIGURE 1.1   Centrepiece in a Paradise Garden, Marrakesh.

Source: Photograph by Clive Hollin.

An ancient garden layout dating to the sixth century BCE has been identified at Passargadae in Persia (now Iran): as with the paradise garden, this design includes four quadrants divided by waterways or pathways. The Persians were expert horticulturists with aqueducts, routed underground to avoid evaporation, to water their fruit trees, shrubs and flowers. In neighbouring Iraq, the Assyrians – who ruled a great empire (900–612 BCE) – similarly took pleasure in gardens, again making creative use of water canals to irrigate rows of trees, vines and flowers. The ancient region of Mesopotamia, which included Assyria and Babylonia, was the location for the development of courtyard gardens and hanging gardens (Dalley, 1993).

In Japan during the Muromachi period (1336–1573 CE) stylised gardens were created at the temples of Zen Buddhism at Kyoto. This type of garden, sometimes called a *zen garden*, sought to replicate the essence of nature, not duplicate its exact appearance, to assist the monks in their meditations. There are three essential elements in a Japanese garden: (i) stone to give structure to the landscape; water, which represents a life-giving force; and (iii) plants to provide changing colour as the seasons change. A zen garden is quite small, surrounded by a wall, and intended to be viewed by a seated observer positioned outside the garden.

The next innovation was to apply botanical knowledge to widen the rewards a garden could provide. The purpose of the *physic garden* was to facilitate the academic study of medicinal plants. The first physic garden was created in 1543 at the University of Pisa, followed by other Italian universities at Padova (in 1545), Firenze (in 1545) and Bologna (in 1547). In the sixteenth century, medicinal gardens spread to other universities, such as Cologne and Prague, and across central Europe. The healing spirit of the physic garden lives on today in the therapeutic use of gardens in hospitals and in our own gardens.

The association between academia and the medicinal garden was evident in that a university appointment as a professor of botany was typically combined with the role of director of the gardens, themselves an adjunct to the Faculty of Medicine. As interest in plants widened, the medicinal garden evolved into the Botanical Garden, the first of which opened in 1544 in the Italian city of Pisa. The idea of a botanical garden took hold in Italy, with gardens established in 1545 in Padua and Florence, and in 1568 in Bologna. The following decades saw cities outside Italy adopt this new type of garden as seen, for example, in Valencia (in 1567), Montpellier (in 1593), Oxford (in 1621), Paris (in 1635), Edinburgh (in 1670) and Amsterdam (in 1682).

In 1772 in London, the merging of the royal estates and plant collections of Richmond and Kew led in 1840 to Kew becoming a national botanical garden. Kew Garden grew to its modern-day size of 300 acres, adding many distinctive features, including the Alpine House, the Nash Conservatory, and the

Palm House. Kew Gardens is one of England's most popular attractions, with over 2 million visitors each year. In 2008 Botanic Gardens Conservation International, based at Kew, London, linked with 800 botanic gardens in 118 countries: there are estimates of 1,775 botanic gardens and arboreta in 148 countries. The contemporary Botanic Gardens performs a range of functions, such as collecting, growing, researching and conserving plants that may otherwise be lost (Westwood, Cavender, Meyer, & Smith, 2021).

In Europe the first gardens were found in Crete, mainland Greece, Sicily and Italy. The ancient Greeks worked the land around 7000 BCE and circa 500 BCE used horticulture for decorative purposes. The first gardens in the sense of an enclosed and planted piece of land alongside a house may well have been in Italy, as seen at Pompeii. These gardens served a practical purpose in proving herbs and vegetables. However, as societies evolved an elite class emerged with the time to appreciate the garden's floral attractions. The servants, sometimes slaves, carried out the hard work of maintaining the garden. The Romans made gardens to complement their palaces and villas. The Roman garden was laid out with hedges and vines and was typically decorated with statues, topiary and a wide variety of flowers, including acanthus, cornflowers, iris and poppies.

In the seventh century CE, the Arab nation ruled a huge empire, and after conquering Persia and Spain, they adopted several ideas about garden design. In particular their gardens used enclosing walls with the plot divided into quadrants by watercourses with a pool in the centre; a variety of trees and flowers completed the design. In medieval Europe by the late thirteenth century CE onwards, gardens, often walled, were grown for secluded pleasure and to produce plants for medicinal use (Landsberg, 2003).

The British Garden

In Britain, as charted by Uglow (2004), the history of gardens and gardening, as opposed to farming, can be traced to the arrival of the Romans in 54 BCE. The Romans brought with them garden designs familiar in the Mediterranean, growing a variety of fruit and vegetables. As the centuries unrolled, gardens became a part of the British landscape inhabited by the aristocracy, nobility and elite of society (Edwards, 2018). The holy populations of the priories and abbeys that dotted the medieval landscape up to 1500 CE required their daily bread, along with copious supplies of fruit and vegetables. These ecclesiastical communities developed systems to cultivate a range of crops, from herbs and salad ingredients to flowers for the bees and their honey, alongside orchards and cornfields. The accompanying ponds were for a supply of fish rather than any ornamental purpose. Glimpses of these, often redesigned, medieval gardens can be seen across England as, for example, at Fountain Abbey in Yorkshire and Avebury Manor in Wiltshire.

Garden Designers

As gardens increased in popularity with the landed gentry, their design assumed increasing importance. A garden design had to reflect the owner's good taste, be interesting to observe, and display contemporary fashion. As in any professional activity, some garden designers rose to prominence within their era. These designers possessed creative ability: they would need to be aware of current fashions, including designs from other countries; they had to possess the technical skills to design buildings and structures; and they had to have the artistic ability to use colour and shape effectively within a given landscape. They could also harness their creativity to give a bespoke flavour to each commission. A brief overview of the notable figures in garden design, ordered by year of birth, is given below.

The Scottish designer William Bruce (c. 1630–1710) took the step of lowering garden walls, allowing an unobstructed view of the surrounding countryside to become part of the garden. This approach is evident at Balcaskie House with a view of Bass Rock and at Kinross House overlooking Loch Leven.

William Kent (c. 1685–1748) started life as a painter, only later becoming an architect and landscaper. He brought about major changes to the layout of estates, for example at Stowe House in Buckinghamshire and Rousham House in Oxfordshire, with a natural style of gardening.

Charles Bridgeman (1690–1738) was an English garden designer who helped pioneer a naturalistic landscape style. He was a key figure in the transition of English garden design away from the Anglo-Dutch formality of patterned parterres and avenues to a freer style that incorporated formal, structural and wilderness elements.

Lancelot Brown (1716–1783), universally known as Lancelot Capability Brown, is the most famous of all the English garden designers. As a young man he was an undergardener with William Kent at Stowe House, at age 26, becoming Head Gardener in 1742. He remained at Stowe until 1750. Brown was permitted by his employer to take freelance commissions, and as his fame grew, he was much sought after. Brown's naturalistic style, called a "gardenless" approach to landscape gardening, replaced the formal, patterned English garden with rolling lawns that ran directly to the house, scattered groves of trees and meandering lakes. Brown designed gardens for around 170 country houses and estates: the list includes Belvoir Castle in Leicestershire, Croome Court in Worcestershire, Blenheim Palace in Oxfordshire, Harewood House in Yorkshire, Warwick Castle and Wynnstay near Wrexham (Figure 1.2).

FIGURE 1.2 Capability Brown.

Source: Photographer Unknown. Creative Commons Licence.

It is a fact of life that anyone bringing about change will be subject to criticism, particularly an astute businessman such as Brown. Wild (2013) describes how Brown's business acumen and social networking led him to become the leading landscaper in the second half of the eighteenth century, with a substantial proportion of the House of Lords among his clients. Brown faced strong criticism from those designers who favoured a structured approach: they declared that Brown persuaded his wealthy benefactors to destroy their inspirational formal gardens with simplistic arrangements of grass, trees and shapeless pools and lakes. The debate about the merits or otherwise of the designs of Launcelot Capability Brown continue to this day (Cox, 2016).

Humphry Repton (1752–1818) offered his services as an "improver of the landscape," and he coined the term *Landscape Gardener*. He used his skills as a watercolour painter to produce his *Red Books* as a shrewd marketing ploy. He first showed the client a page of the book with a painting of the garden as it stood then. Raising a flap revealed a magnificent new vision of the garden, specifically designed to please the client. These red leather-bound books gave intricate details for the proposed changes, including drawings, maps, plans, and before-and-after sketches and watercolours; some of these Red Books survive in several collections. Repton's style was influenced by Capability Brown with an emphasis on the house and how it was placed within the surrounding landscape. Repton saw gardening as an art form with its natural beauty enhanced by features such as terracing, gravel walks and flower beds. However, unlike Brown, Repton, left it to the client to commission the labour force to turn his plans into reality.

Repton expounded his formula for landscape gardening in his 1975 book *Sketches and Hints on Landscape Gardening,* followed in 1803 by *Observations on the Theory and Practice of Landscape Gardening.* In his career he accepted over 400 commissions from owners of stately homes – including Harewood House in Yorkshire, Longleat in Wiltshire, Sheringham Park in Norfolk, Tatton Park in Cheshire and Woburn Abbey in Bedfordshire – where his designs can be seen.

Although his name has not been as enduring as Capability Brown or Humphry Repton, the Victorian Scottish designer John Claudius Loudon (1783–1843) can be remembered for three principal contributions to the history of gardening. First, he felt that gardeners should get their hands dirty and not rely on paid labour, while emphasising the centrality of plants in garden design. Second, he founded the periodical *The Gardener's Magazine* and was a prolific writer, his books included the extensive *Encyclopedia of Gardening* and *Hortus Britannicus,* a catalogue of every plant found in Britain. Third, Louden was an advocate of green spaces and parks in crowded cities for the enjoyment and health of the people, and to this end, in 1839 he created Derby Arboretum (now Grade II listed) where he planted and labelled over 1,000 plants.

Edwards (2018) makes the point that while women were largely absent from the history of gardening up to the Victorian era, "Once the garden was understood as essentially decorative, women could claim it as part of the domestic sphere" (p. 171). This perceptual shift opened the door for women to play a role in garden design, and many came to the fore, some of whom are noted below.

Gertrude Jekyll (1843–1932) designed more than 400 gardens in Great Britain, Europe and America. Her designs were based on a formal structure, relaxed by large groups of colourful plants. This style has been of enduring popularity in Britain, with colour always a prominent feature. The hardy English shrub rose *Rosa Gertrude Jekyll,* with large pink spiral patterned blooms, stands in her memory.

Margery Fish (1892–1969) is best known for the garden, now Grade I listed, she created at East Lambrook Manor in Somerset. Her style of gardening was a product of the times: WWII made labour scarce so that costly gardeners were no longer viable. She designed a large-scale cottage garden with an informal planting style that included perennials and ground cover to save time weeding. She wrote a string of gardening books between 1956 and 1970 alongside regular contributions to gardening magazines.

The novelist and poet Vita Sackville West (1892–1962) and her husband, Harold Nicolson, created the now Grade I listed gardens at Sissinghurst Castle in Kent. They pioneered features such as a multi-layered planting style, colour-themed gardens, particularly the White Garden, and the division of the large plot into distinct smaller "rooms". Sissinghurst has ten such rooms, formed from walls and yew hedges, and the planning draws the eye from the entrance through to the exit with a feature such as topiary or a partly hidden sculpture in the room beyond.

Beth Chatto (1923–2018) transformed her family fruit farm at Elmstead Market near Colchester in Essex into the Grade II listed Beth Chatto Gardens. The land was of variable quality, in some places very dry or too wet, with patches of gravel and thick clay, and heavily shaded areas. This range of conditions could have been seen as problematic, but Chatto turned them to her advantage, creating habitats such as a gravel garden, a shady walk, and a scree garden. Chatto wrote books and articles about planting in these "problem areas" using flora that have developed to grow in these various conditions.

As will be seen immediately below, these designers were influential in their own time. However, as explored further in Chapter 7, their many and varied influences are to be found in the modern-day suburban garden.

The Tudor Period (1486–1604)

The Reformation and the destruction of the monasteries (and their gardens) began in 1536 during the reign of King Henry VIII. It was during the Tudor period that gardens moved away from functional food production with the incorporation of decorative features such as the knot garden, crafted turf mounds, and labyrinths and mazes. Gardens also became a place for recreation: in 1588 Sir Francis Drake is famously said to have been playing bowls while the Spanish Armada sailed up the British Channel. Today, as Edwards (2018) notes, "Generally only scraps of Tudor gardens remain … . There is one astonishing exception – the haunting poetic landscape of Lyveden New Bield" (p. 45).

The Stuart Period (1604–1714)

Following the Tudors, the Stuart period saw new horticultural ideas, particularly from France and Italy, flowing from the opening up of English trade and travel with the continent. Inspired by the French, the *parterre* became popular, displacing the Tudor knot garden: a parterre, meaning *on the ground*, is symmetrical patchwork of enclosed, sometimes with box hedging, adjacent flower beds divided by gravel. The parterre is intended to be seen from a height, such as the upper floor windows of a country house. The continental love of water in the garden was another adopted feature: capitalising on the new understanding of hydraulics, fountains spouted water to great heights, statues gushed into pools, canals flowed and the great house was reflected in a large lake. A note of whimsy was introduced with the Italian *giochi d'acqua* and French *jeu d'eau*: for example, the practical joke of water jets suddenly turned on by a gardener to drench unsuspecting guests was a source of much amusement.

The importation of ideas from abroad was not limited to Europe with Chinese features becoming popular. For example, the garden at Stowe, inspired by William Kent, has a Chinese House in the form of a miniature folly, intricately adorned with Chinese scenes, flora and lettering.

The *National Trust* lists several extant Stuart gardens, including Blickling in Norfolk and Chastleton House in Oxfordshire.

The Georgian Period (1714–1830)

The Georgian period saw a drift away from the garden as a statement of wealth and power to being an expression of self: your garden proclaimed your beliefs and said that you were educated with good taste. A defining aspect of the Georgian garden was its embrace of the search for *Arcadia*, a poetic, idyllic vision of pastoralism and a return to harmony with an unspoilt nature. This informal style of garden layout, championed by Capability Brown, contrasted sharply with the geometric style of the Stuart era. Thus, for example, lakes were created to be in harmony with the landscape, cascading water gave movement and dramatic effect, while buildings such as follies and temples provided sheltered spaces to rest and to take refreshment.

The availability of imported plants prompted the inclusion of previously unseen exotic plants from Australia, China, the East and West Indies, South Africa, South America and Russia. These new arrivals, including species of American maples and oaks, the Maidenhair Tree (*Gingko biloba*), and the Sweetgum Tree (*Liquidambar styraciflua*), along with new species of conifer, gave fresh form and colour to the garden. In addition, there were stunning displays of autumn foliage. Yet further, greenhouses and orangeries made it possible to collect and display an expanding range of tropical and sub-tropical plants.

The desire for gardens to be open spaces with sweeping vistas meant a minimal use of hedges and fencing as they imposed limits on the views. However, removing tangible boundaries created the problem of keeping livestock away from precious plants. The solution took the form of the *ha-ha*, a sunken barrier formed by digging a deep ditch with its inner side built up to ground level with a wall. The outer side slopes steeply upwards before smoothing out at ground level. Thus, from the garden the view is uninterrupted; from the outside the walled ditch sitting below ground level is an effective barrier. First described in 1709 by the gardening enthusiast Dezallier d'Argenville in his *La Theorie et la Practique du Jardiance*, d'Argenville said that the name "ha-ha" originated with the optical illusion engineered by design so that the person discovering the concealed ditch and wall would cry *"Ah! Ah!"*.

Installed by Charles Bridgeman and John Lee in the 1720s, the gardens at Stowe in Buckinghamshire may have been the first in England to possess a ha-ha. This innovation stirred the attention of visitors from the landed classes, who took the idea and incorporated it into their own estates.

The Victorian Period (1837–1901)

The long reign of Queen Victoria spanned a period when Britain became a global superpower, wielding military and industrial power on a vast scale. The British Empire included almost one-quarter of the world's population – Australia and New Zealand, the Caribbean, North America, significant portions of Africa, and the Indian subcontinent – all with Victoria as head of state. The industrial revolution, utilising the power of steam to build fast, powerful machines, prompted an exponential growth in factories, particularly in textile production. On this wave of change, Britain was driven to find new markets for its goods while ensuring easy access to raw materials from foreign countries. The country's wealth funded substantial expenditure on civic pride with significant investment on town halls, libraries, swimming baths and other public amenities.

In keeping with Britain's global expansion, alongside familiar favourites, Victorian gardens showcased plants garnered from many countries: flowerbeds blazed with colour and arboreta displayed strange trees such as the Monkey Puzzle Tree, while fruit trees also became fashionable. The design of the Victorian garden emphasised symmetrical tidiness with flowerbeds alongside rockeries, inspired by reports from explorers of mountainous areas on the other side of the globe.

Kitchen gardens, often walled, grew a wide range of fruit and vegetables, capitalising on advances in glass house design and construction. The Palm House at Kew Gardens, originally built in the 1840s, used construction techniques employing wrought iron, which made it possible to build large glasshouses. Ferns, a particular Victorian favourite, and tender, exotic plants such as pineapples prospered in glasshouses. On a smaller scale greenhouses became popular, and arched wrought iron frames were used around the garden to support climbers.

The Victorians were fond of embellishments, such as birdbaths, sundials, statues and urns, which were either placed discreetly around the garden or boldly in a prominent position as a focal point. A water feature such as a fountain added sound and movement to the Victorian ambience. The widespread availability of garden ornaments owed much to Mrs. Eleanor Coade's invention of *Coade Stone*. This substance was not actually stone; rather, it is a ceramic formed from a mixture of clay, glass silicates and terracotta, which is fired for four days in extremely hot kilns. Coade Stone was flexible but strong, and the Victorians used it as a substitute for real stone in designing architectural features, monuments, sculptures and garden ornaments.

There are several Victorian gardens to be found in Britain: *Cragside* in Northumberland has one of the largest rock gardens in Europe, and flower beds are planted vivid summer colours. Cragside also has a collection of North American conifers, while an orchard house allows the cultivation of early fruit. *Biddulph Grange* in Staffordshire is designed as a series of inter-connected rooms, each given to distinct parts of the world, including China with a bright red pavilion, Egypt and its pyramids, and a Himalayan glen.

The next period moves to the contemporary garden, from the Edwardian period to the present day.

As the Victorian era gave way to the Edwardian, garden design gave a sigh of relaxation, and Victorian formality gave way to naturalistic planting, albeit within a structured design. The Edwardian garden is distinguished by hedges, straight paths, terraces and sunken gardens. The rejection of mass production in favour of traditional artistry was a tenet of the emergent Arts and Crafts movement inspired by John Ruskin (1819–1900) and William Morris (1834–1896). The influence of Arts and Crafts on garden design was apparent in handmade stonework, such as stone-flagged terraces and dry-stone walls, alongside topiary, hedges, pleached trees to create the effect of a green wall, orchards and rose beds. Sissinghurst Castle in Kent is a fine example of a garden from this period.

As the 1800s drew to a close, British society faced the societal problems caused by industrial expansion. The social reformer Ebenezer Howard (1850–1928) argued for a new type of town separating industrial from residential areas. This design provided the advantages of city and country-side while eradicating their disadvantages. The first of these new conurba-tions, built in 1904, was called Letchworth Garden City (Miller, 2002), closely followed by Welwyn Garden City in Hertfordshire, Brentham Garden Suburb in Ealing, London and Hampstead Garden Suburb in London. The garden city became a global phenomenon, with examples to be found in Africa, Asia, Australia, Europe, New Zealand, and North and South America.

Garden cities have been criticised as economically damaging, destroying nature and inconvenient for travelling in and out of.

References

Cox, O. (2016). Why celebrate Capability Brown? Responses and reactions to Lancelot 'Capability' Brown, 1930–2016. *Garden History*, 44(1), 181–190.

Dalley, S. (1993). Ancient Mesopotamian gardens and the identification of the hanging gardens of Babylon resolved. *Garden History*, 21(1), 1–13.

Edwards, A. (2018). *The story of the English garden*. London: National Trust Books.

Landsberg, S. (2003). *The medieval garden*. Ontario, Canada: University of Toronto Press.

Miller, M. (2002). *Letchworth: The first garden city* (2nd ed.). Chichester: Phillimore.

Spencer, R., & Cross, R. (2017). The origins of botanic gardens and their relation to plant science with special reference to horticultural botany and cultivated plant taxonomy. *Muelleria, 35*, 43–93.

Uglow, J. (2004). *A little history of British* gardening. London: Chatto & Windus.

Westwood, M., Cavender, N., Meyer, A., & Smith, P. (2021). Botanic garden solutions to the plant extinction crisis. *Plants, People, Planet, 3*(1), 22–32.

Wild, A. M. (2013). Capability Brown, the aristocracy, and the cultivation of the eighteenth-century British landscaping industry. *Enterprise & Society, 14*(2), 237–270.

2

GARDENS FOR ALL

For centuries war has changed the course of human history (LeBlanc & Register, 2003), bringing about significant, sometimes unanticipated, national changes. In England and Wales, the effects of war on the humble garden may be seen in more recent times. The period between the two world wars was marked by an expansion in house building, creating millions of new houses, many with a garden (Constantine, 1981; Hollow, 2011). Thus, in the short inter-war period the number of people owning a private garden reached a new high, and while not every new house owner became a gardener, many did. There were yet more new houses and gardens in the post-war housing boom of the 1950s, adding to the number of people for whom the suburban garden became a part of their everyday life (Langhamer, 2005). It seems unlikely that those responsible for housing developments would have anticipated the social and economic consequences of the formation of thousands of small-scale gardens. Indeed, the average garden measures about 150 square feet, minute by comparison to its predecessors, but they all add up; as Richards and Farrell (2020) observe, "In the UK alone, private gardens occupy an area one-fifth the size of Wales" (p. 10). For those unfamiliar with Wales as a unit of area, this equates to around 1,605 square miles of garden within the United Kingdom. Although allotments have a long history (see Chapter 4), it was the rapid and widespread exposure of large swathes of the population to the rewards of gardening that brought about several consequences: first, the identification of new commercial opportunities; second, an expansion of gardening media; and third, a new setting for social interactions.

DOI: 10.4324/9781003289661-3

Commercial Opportunities

The garden nursery was traditionally the place where people went to buy plants. When there were a small number of gardens, these were specialist establishments with skilled and knowledgeable staff. However, the growth in the number of gardens opened up new possibilities for selling plants to the public. In the early 1950s, Edward Stewart, a keen horticulturalist from southern England, took a trip to Canada, where he encountered a new way of selling plants – the Garden Centre. On his return in 1955, he brought home the idea and established the Stewarts Garden Centre in Ferndown, Dorset. Since then, the numbers of garden centres have grown exponentially. The value of the UK garden retail market, with over 2,500 outlets, runs to billions of pounds per annum. Hamilton (2005) observes that:

> The term 'garden centre' can, in effect, mean many different things. At the top of the market there are the state-of-the-art centres merchandising a wide range of products – they are both specialist retailers of garden plants and other associated products, and more general retail centres in their own right. In the middle market there are smaller centres with site areas of two to four acres, and stores specialising in DIY retail that have garden centres attached. Finally, at the other end of the scale, are the small, horticulturally based nursery businesses that have expanded into the retail sale of their home-grown plants and associated garden products. (p. 32)

Garden centres present the gardener with an ever-widening range of choice, not just in varieties of plants but also in a range of garden accoutrements such as tools and ornaments. How gardeners responded to this choice, later enhanced by online retailers, in shaping their individual plots is considered in Chapter 5.

The increased number of gardens presented financially rewarding opportunities for both jobbing and specialist gardeners. If you liked your garden but were not so keen on the hard work, you could always hire a gardener to do the digging; alternatively, you could hire a specialist to tend your lawn or prune your trees.

Gardening Media

We are surrounded by the written and electronic media in its multiplicity of forms. There is no doubt that our behaviour is influenced, for the better or the worse, by what we see, hear and read.

Gardens in Print

William Turner (1508–1568), sometimes referred to as the father of English botany, published *The New Herbal* in 1551, followed by the second part in 1562 and the third in 1558. For the first-time these three books presented listings written in English rather than Latin, which allowed people easily to identify the main English plants. Turner did not give any advice on growing or caring for plants. In 1563, Thomas Hill wrote *A Most Briefe and Pleasaunte Treatise, Teaching How to Dresse, Sowe, and Set a Garden*, the first book about gardening published in England. Hill also wrote *The Profitable Art of Gardening* published in 1568, which discussed designs for the then popular knot garden. At the time of their publication, the potential readership for these books was small. However, publications advanced between the wars, so magazines reached a much larger audience (Constantine, 1981). Thus, the up-market *Homes and Gardens* sat on the newsagent shelves alongside the more prosaic *Amateur Gardening*, *Gardner's World*, *Garden News* … and the list goes on.

Gardens on Air

Public broadcasting, radio and television cater to people's interests and, as in other areas of life (Stever, 2019), gardening programmes have thrown up a raft of celebrities (Figure 2.1). Over the years these gardening luminaries have included, among others, Percy Thrower, Geoff Hamilton, Alan Titchmarch and Monty Don. These media personalities have been admired by fellow gardeners, and some have become celebrities in their own right. Onu, Kessler and Smith (2016) make the point that when we admire another person this can influence our own behaviour. We know from social learning theory that people will

FIGURE 2.1 A Well-Known TV Gardening Celebrity.

Source: Picture © Andy Mabbett (pigsonthewing.org.uk), under CC by-sa 3.0 licence.

imitate those they perceive to be high status role models. Thus, we may learn new skills and techniques as well as gain knowledge from celebrity gardeners (Meng-Lewis et al., 2021). In addition, through the celebrity we may feel connected to a wider community, seeking the rewards of contact, cooperation and assistance from those who hold similar interests to our own.

Gardens on the Internet

In the digital age, we can access a vast range of material, from home and abroad, via podcasts and web sites. Some websites, such as that of *The Royal Horticultural Society*, have a staggering range of information covering types of plants, gardening events and merchandise available for purchase. The extent of the detailed information available to anyone with access to the Internet is unprecedented in the history of gardening.

Social Opportunities

As gardening became more widespread, it created new opportunities for the rewards inherent in social contact with other people. Although some people were happy to work in isolation, others saw residential gardens as a chance for social interaction and more. *The National Garden Scheme* (NGS) was created in 1927 with the aim of "opening gardens of quality, character and interest to the public for charity" in England, Northern Ireland, Wales and The Channel Islands. The first NGS event saw 609 private gardens opened to the public, raising £8,191 for charity. Over the decades the NGS has grown such that over 3,500 gardens are now opened annually (the events take place on specific dates for each village or town) with over £60 million raised for worthy causes. These events are social occasions as people mingle and discuss the finer points with exhibitors and fellow visitors.

As well as private gardens going on show, there are national flower shows, including the *Chelsea Flower Show, Tatton Park Show* and *RHS Hampton Court Garden Festival,* which is the largest flower show in the world. There are many large gardens (often with a stately home) that are open most of the year to the public. These shows and gardens offer a variety of opportunities, such as meeting other gardeners, seeking advice from experts, viewing new plant varieties, and perusing the inevitable retail outlets. This is all big business, as Connell (2004) makes clear:

> Gardens play a significant role in the enjoyment of leisure time and the pleasures derived from the garden environment extend well beyond the parameters of the domestic garden, with some 16 million visits made to British gardens open to the public every year. (p. 229)

A Place for Growing

Alongside the flora, the garden is a place for children to grow and develop. A garden can provide the child with places – especially secret places (Moore, 2014) – for adventure with trees and tree houses to climb, fruit and vegetables to nibble, lawns for games, shrubberies to disappear into, and water features to watch wildlife. There may be garden toys, such as paddling pools and trampolines, for solitary play or with friends. Laaksoharju, Rappe and Kaivola (2012) suggest that alongside play and social learning with friends and family, gardens also promote the child's relationship with nature.

Contact with nature is widely seen as advantageous for the child's health and well-being (Chawla, 2015; Mustapa, Malikia, & Hamzah, 2018). Chawla presents a lengthy list of advantages for the child who has access to nature; these benefits range through physical and psychological health, individual and social development, and promoting healthy eating. The idea underpinning the child's purported need for nature is called the *biophilia hypothesis*. Hand et al. (2017) explain: "The biophilia hypothesis proposes that humans have an innate tendency to affiliate toward life and life-like processes as a consequence of evolution, where survival and reproduction were dependent on interactions with the natural environment" (p. 274). The logical extension of this position is that children should be afforded as much contact with nature as possible. In turn, this point is reinforced by the growing disconnect of children with nature as greater amounts of time are spent with electronic entertainments.

Of course, it is desirable that children are raised in a physically and socially safe and supportive environment that nurtures their growth (Christian et al., 2015). The child's putative need for nature can be met by greening the environment with trees and parks and by visits to wildlife settings. The inclusion of gardening in the school curriculum can bring an interaction with nature to a large number of children (Earl & Thomson, 2020; Huelskamp, 2018; Taylor, Wright & O'Flynn, 2021). However, Wake (2008) cautions against the "You know what you want, I know what you need" approach in which children's gardens are planned by adults from their own perspective rather than the child's. From a wider viewpoint Cairns (2018) warns of a "magic carrot" by which "children's garden encounters will magically reshape society" (p. 517). This point will be considered in Chapter 4 when discussing gardens in schools.

Efforts to connect children with a wider, green world may pay dividends in the child's later life, particularly with regard to acquiring a respect for the natural world. Lohr and Pearson-Mims (2005) suggest that a child's interactions with plants – in which parents can play a significant role (Fančovičová & Prokop, 2010; Remmele & Lindemann-Matthies, 2018) – lead to positive attitudes towards greenery in later life. It is not stretching a point to suggest that such positive attitudes, as seen in the English city of Sheffield (Heydon, 2020), may engender strong and active support for the protection of urban trees (Jones, Davis, & Bradford, 2012), not the least because of the association with house prices (Orland, Vining, & Ebreo, 1992).

The Green Gardener

There is no doubt that we are living in a time of perilous climate change. Indeed, the suggestion has been made that we are entering a new period in our planet's history, the *Anthropocene* (Crutzen & Stoermer, 2000; Lewis & Maslin, 2015). In the Anthropocene – *anthropo* meaning man, *cene* meaning new – humans are a major destructive ecological and geological force bringing unspoiled nature close to an end such that Earth's systems that support life are at grave risk of irreparable damage. It may seem a forlorn hope that the individual gardener can save the planet, but there are lots of gardeners with many gardens, so perhaps some compensation is possible? As Goddard, Doughill and Benton (2013) state: "Private gardens are a major component of cities in both developed and developing world countries ... and the manner in which householders manage these spaces has a substantial impact on the provision of urban biodiversity" (p. 258). Davies et al. (2009) conducted an analysis of archival data of domestic gardens across the United Kingdom, suggesting that 87% of households, about 22.7 million homes, have access to a domestic garden. They considered the ways in which gardeners provide resources to aid biodiversity: a great deal of effort was given to helping birds (see below) through the provision of food and nest boxes, alongside maintaining ponds and planting trees.

Goddard, Doughill and Benton (2013) looked at gardening for wildlife in a household questionnaire survey carried out in Leeds, England. The survey elicited 563 responses with the main findings that about three-quarters of households provided bird food in their garden, the majority continuing to provide food during winter. The gardeners revealed that their main rewards for wildlife-friendly gardening were twofold: (i) a sense of personal well-being; and (ii) fulfilling a moral responsibility to nature. With regard to the first reward, the respondents said they felt personal satisfaction at their success in bringing wildlife to their garden, alongside the calm and peacefulness associated with watching the birds. The second reward involved a sense of taking a step towards bridging the disconnect between urban living and nature, thereby helping to conserve wildlife.

A barrier to gardening for wildlife entails allowing plants to flourish to attract insects and other fauna. Unmown lawns, uncut hedges and letting weeds flourish results in an untidy garden, particularly noticeable with a garden at the front of the house, which may run counter to neighbourhood standards. Goddard, Doughill and Benton found that some gardeners did not, understandably, want to contravene local social norms and toned down the extent of their wildlife-friendly pursuits. As well as local standards, another barrier to wildlife-friendly gardening lay in a lack of information or knowledge as to how best to proceed. van Heezik, Dickinson and Freeman (2012) make the point that neighbourhood standards and a lack of information need not be barriers to wildlife gardening. They worked with 55 householders to increase their knowledge of native biodiversity and environmentally friendly gardening techniques. There was both a growth in knowledge and a change in attitudes, leading just over one-quarter of the householders to make positive changes to their gardens. van Heezik, Dickinson and Freeman note the potential for the wildlife-friendly changes in gardening to influence the behaviour of neighbouring gardeners and so shift local standards.

The Birds, the Bees, the Other Fauna

The Davies et al. study cited above reported that in about one-half of the gardens there was provision for supplementary food for birds, with about half of the sample using bird feeders. They estimate that these figures indicate "A minimum of 4.7 million nest boxes within domestic gardens" (p. 767). Why do gardeners feed the birds? What rewards this behaviour?

Goddard, Dougill and Benton (2013) suggested that gardening for wildlife can help the gardener's general well-being as well provide a connection with nature.

Clark, Jones, and Reynolds (2019) conducted an online survey of 563 people, 62% female, 38% male, asking what led them to feed the birds and what led to and rewarded this activity. The three most often cited antecedents were the influence of parents (57%), having a garden (53%) and grandparents (43%), while the top five rewards for feeding were pleasure, helping birds to survive, nurture, children's education and a connection with nature. Thus, bird feeding brings the gardener the external reward of the presence of birds and internal rewards of pleasure and a sense of helping fellow creatures. The strength of these rewards may be mitigated by the bird species; it seems we all love a robin but a woodpigeon less so (Cox & Gaston, 2015).

The bee is particularly valuable for the gardener in its role as a pollinator – no bees, no flowers – so it is to everyone's advantage that, alongside other pollinators, bumblebees, honeybees and solitary bees flourish in our green spaces. There are several strategies gardeners may use to attract pollinators (Griffiths-Lee, Nicholls, & Goulson, 2022; Lanner et al., 2020; Rahimi, Barghjelveh, & Dong, 2022): (1) judicious planting and adding wildflowers to the traditional favourites to extend the flowering season from March to October; (2) developing microhabitats such as sparsely vegetated patches of ground; and (3) creating artificial nesting sites with "bee hotels". However, it cannot be assumed that all plants suit all pollinators; for example, different types of bees need different plants (Anderson et al., 2020).

Given the centrality of pollination to gardening, and in the face of what has been called an "insect apocalypse" (Goulson, 2019), why do more gardeners not employ the strategies above to attract pollinators to their plots? As Drivdal and van der Sluijs (2021) note, it may be that the lack of global action and clear national policies regarding the pollinator crisis forestalls gardeners from acting. Drivdal and van der Sluijs suggest that given this situation it may be best to adopt the Precautionary Principle (PP), which

> Means that when it is scientifically plausible that human activities may lead to morally unacceptable harm, actions shall be taken to avoid or diminish that harm … . The application of the PP does not prescribe a particular course of action: varying degrees of precaution may be taken, from strong to weak. (p. 96)

They give two examples of this precautionary approach: (1) effective regulation of *neonicotinoids,* which are neurotoxic insecticides that cause severe damage to beneficial insects such as pollinators; and (2) curtailment of the international commercial trade in honeybees and bumblebees, which is both a threat to local biodiversity and further endangers the decline of pollinators by the spread of parasites.

A precautionary approach is based on taking steps to circumvent an undesirable consequence of our actions and therefore relies on the design of contingencies based on *negative reinforcement*: i.e., people's behaviour changing to *avoid* an unwanted outcome. The immediate problem is one of scale: changing the widespread use of insecticides in farming and altering patterns of commercial trade is radically different from changing the behaviour of the individual gardener. Nonetheless, the gardener has an important role to play: given the acreage involved, if enough gardeners take positive action, then perhaps all is not lost. What stands in the way of affirmative action, and how may barriers be overcome?

Hall and Martins (2020) suggest that while there is a growing awareness of the decline in bees, along with other stinging insects the bees themselves have an image problem; they are perceived "as a conditional danger – dangerous if provoked" (p. 108). Hall and Martins also note the absence of co-ordinated policies for preserving pollinators and suggest that public education, both formal and informal, such as providing information at garden centres, may shift perceptions away from insects as dangerous and support pollinator-friendly gardening. Indeed, while the need for pollinator-friendly gardening is generally appreciated, the gap between knowledge and action shows that gardeners have much to learn about their role in conservation (Lindemann-Matthies, Mulyk, & Remmele, 2020; Wilson, Forister, & Carril, 2017). As well as horticultural information, efforts to change gardeners' behaviour could capitalise on the personal rewards associated with pollinator conservation, including a connectedness with nature and pleasure at contributing successfully to bee conservation (Knapp et al., 2021; Sturm et al. 2021).

Of course, birds and bees are not the only fauna that visit a garden. A survey of residential gardens in western Massachusetts, using camera trapping technology, found that the gardens were visited by 16 non-domesticated species, including black bears, chipmunks, deer, foxes, opossum and racoons (Grade, Warren, & Lerman, 2022). In the United Kingdom there is an absence of bears, but our gardens may become habitats for a range of mammals, including deer, foxes, hedgehogs, mice, rabbits and squirrels. The hedgehog is a welcome garden visitor as its diet includes plant-destroying slugs and snails (having said which, it may also take birds' eggs and nestlings). The provision of suitable food is one of the most effective ways to attract hedgehogs into the garden (Gazzard, Yarnell, & Baker, 2022).

Hedgehogs have voracious appetites and may visit twenty or more gardens a night as they forage for food. Their ability to move depends on accessible routes between gardens; however, the use of wooden fences, particularly those with gravel boards, reduces connectivity between gardens. The "Hedgehog Street" public engagement campaign recruits volunteers to make access points across gardens to improve connectivity. Gazzard et al. (2021) reported that the campaign had met with some success, although it stalled when garden owners had concerns about boundary ownership or about negotiating with neighbours on how to proceed.

Utilising Natural Resources

The green gardener will wish to utilise natural resources such as composting garden waste (which helps attract wildlife, including hedgehogs) and conserving water. It is understood that there are times when we must conserve water, but this can pose a problem for the gardener. Many but not all gardeners have rainwater butts; those without use hosepipes and sprinklers to keep lawns and plants alive. Chaudhary, Lamm and Warner (2018) use the psychological principle of *cognitive dissonance* to encapsulate the gardener's watering dilemma. When we have two conflicting thoughts – "I agree with water conservation" versus "I must water my garden when it needs it" – we are in an uncomfortable psychological state of dissonance. We may attempt to reduce dissonance by either changing our beliefs one way or the other or by adopting new behaviours (e.g., installing a water butt).

Looking Back

How does the contemporary garden compare with its predecessors? The modern garden is significantly smaller than before, and although the basic design elements remain unchanged, there is now considerably less space for a much greater choice of trees, shrubs, flowers and accessories. There are greenhouses and cold frames to suit all sizes of gardens and sheds to provide the gardener with a place to potter in quiet contemplation. Garden ornaments, say a fox or a ladybird, may add a quirky interest, and some gardeners may inject a spot of humour with a garden gnome or two. Water features remain popular, while garden ponds and fishkeeping have risen to new levels of technical sophistication. The garden remains a place for socialising and eating, although who can say what the Victorians would make of modern BBQ equipment!

One particular aspect of garden design that has taken on a new guise relies on the use of a visual illusion (Gregory, 1997). As discussed previously, the ha-ha was devised to create the illusion that there was no disconnect between the boundaries of the stately garden and the adjoining countryside. Optical illusions have been appreciated for centuries, Aristotle described the *waterfall illusion* whereby after staring at a waterfall for a little while objects close by appear to be moving upwards. There are many visual illusions: as shown in Figure 2.2, the Muller-Lyer Illusion where two lines of equal length appear unequal because of the direction of the arrows, is a perennial textbook favourite (Woloszyn, 2010).

There is a great deal of research and theorising given to why we perceive visual illusions (Shapiro & Todorovic, 2016). The garden provides a perfect setting for illusions to be put to work. An illusion can give a sense of greater space; for example, the reflections in a judiciously situated mirror give the impression of the garden stretching out beyond. A mirror also bounces light into dark corners, giving the illusion of depth and space beyond the wall on which it hangs (Figure 2.3).

A similar illusion of space can be achieved by using hedges to divide a long thin garden into a half-hidden sequence of small plots, seemingly stretching off into the distance. The same effect can be achieved by shaping paths to curve around bends and corners so that they appear to lead off to pastures new.

FIGURE 2.2 The Muller-Lyer Illusion.

Source: This Photo by Unknown Author is licensed under CC BY-SA-NC.

FIGURE 2.3 Garden Mirror.

Source: Photographer Larisa Birta.

Looking Forward

What innovations might future generations ascribe to our era? At a time of climate change it is inevitable that environmental concerns will have a profound effect on gardeners and their gardens. Beth Chatto's approach to gardening may become increasingly relevant as we create green buildings, their walls and roofs a burst of greenery in an attempt to overcome the "urban heat island effect" in the concrete city. Gardeners will grapple with the effects of climate change and the changing possibilities presented by growing plant species, such as tree ferns and bananas, which thrive in warm, wetter weather. Environmental concerns may increase the numbers of ponds and bird feeders and prompt the introduction of wildflowers and other insect-friendly plants. These changes, in turn, offer the gardener the intrinsic rewards to be found in feeding the fish, watching bees taking nectar, and seeing different varieties of birds visiting the feeders. Yet further, Cameron et al. (2012) describe how gardening practices can, alongside parks and other green spaces, support the planet by improving local air cooling, lowering the chances of flooding (especially when people resist turning their front gardens into a car park), and provide a safe place for wildlife. In addition, not introducing invasive alien plant species such as knotweed (Bailey, 2011) and not using pesticides and noxious weedkillers is of clear benefit.

Way (2020) notes that the garden at the front of the house was customarily looked after to enhance the appearance of the house and provide a source of interest for passers-by. Indeed, Way cites a book by John Claudius Louden published in 1812, which provides 26 designs for the suburban front garden. The traditional front garden, bisected by the path to the front door, displayed a lawn and a tree, with colour provided by roses and a myriad of other herbaceous plants, including lupins to give height, completed by a vegetable garden.

Although the garden is a source of floral reward, it offers much more. We are encouraged to think of our garden as another room, an alfresco extension to our house, where we can continue doing outdoors what we do indoors. So, we cook, eat and drink in the garden, children play and keep their pets, we meet with family and friends, and host parties. Thus, history has given a framework for considering the rewards contemporary gardening offers, and three topics may be identified: (1) the beauty of a garden; (2) socialisation in the garden; and (3) the individual gardener. What might psychology bring to understanding more about these three areas?

References

Anderson, H. B., Robinson, A., Siddharthan, A., Sharma, N., Bostock, H., Salisbury, A., ... & Van der Wal, R. (2020). Citizen science data reveals the need for keeping garden plant recommendations up-to-date to help pollinators. *Scientific Reports*, *10*(1), 1–8.

Bailey, J. (2011). The rise and fall of Japanese knotweed? In I. D. Rotherham and R. A. Lambert (Eds.), *Invasive and introduced plants and animals – Human perceptions, attitudes and approaches to management* (pp. 221–232). London: Earthscan.

Cairns, K. (2018). Beyond magic carrots: Garden pedagogies and the rhetoric of effects. *Harvard Educational Review*, *88*(4), 516–537.

Cameron, R. W., Blanuša, T., Taylor, J. E., Salisbury, A., Halstead, A. J., Henricot, B., & Thompson, K. (2012). The domestic garden – Its contribution to urban green infrastructure. *Urban Forestry & Urban Greening*, *11*(2), 129–137.

Chaudhary, A. K., Lamm, A. J., & Warner, L. A. (2018). Using cognitive dissonance to theoretically explain water conservation intentions. *Journal of Agricultural Education, 59*(4), 194–210.

Chawla, L. (2015). Benefits of nature contact for children. *Journal of Planning Literature, 30*(4), 433–452.

Christian, H., Zubrick, S. R., Foster, S., Giles-Corti, B., Bull, F., Wood, L., ... & Boruff, B. (2015). The influence of the neighborhood physical environment on early child health and development: A review and call for research. *Health & Place, 33*, 25–36.

Clark, D. N., Jones, D. N., & Reynolds, S. J. (2019). Exploring the motivations for garden bird feeding in south-east England. *Ecology and Society, 24*(1), Article 26.

Connell, J. (2004). The purest of human pleasures: The characteristics and motivations of garden visitors in Great Britain. *Tourism Management, 25*(2), 229–247.

Constantine, S. (1981). Amateur gardening and popular recreation in the 19th and 20th centuries. *Journal of Social History, 14*(3), 387–406.

Cox, D. T. C. & Gaston, K. J. (2015). Likeability of garden birds: Importance of species knowledge & richness in connecting people to nature. *PLoS One, 10*(11), e0141505.

Crutzen, P. J., & Stoermer, E. F. (2000). The 'Anthropocene'. *Global Change Newsletter. The International Geosphere-Biosphere Programme (IGBP): A Study of Global Change of the International Council for Science (ICSU), 41*, 17–18.

Davies, Z. G., Fuller, R. A., Loram, A., Irvine, K. N., Sims, V., & Gaston, K. J. (2009). A national scale inventory of resource provision for biodiversity within domestic gardens. *Biological Conservation, 142*(4), 761–771.

Drivdal, L., & van der Sluijs, J. P. (2021). Pollinator conservation requires a stronger and broader application of the precautionary principle. *Current Opinion in Insect Science, 46*, 95–105.

Earl, L., & Thomson, P. (2020). *Why garden in schools?* London: Routledge.

Fančovičová, J., & Prokop, P. (2010). Development and initial psychometric assessment of the plant attitude questionnaire. *Journal of Science Education and Technology, 19*(5), 415–421.

Gazzard, A., Boushall, A., Brand, E., & Baker, P. J. (2021). An assessment of a conservation strategy to increase garden connectivity for hedgehogs that requires cooperation between immediate neighbours: A barrier too far? *PloS One, 16*(11), e0259537.

Gazzard, A., Yarnell, R. W., & Baker, P. J. (2022). Fine-scale habitat selection of a small mammalian urban adapter: The West European hedgehog (*Erinaceus europaeus*). *Mammalian Biology, 102*(2), 387–403.

Goddard, M. A., Dougill, A. J., & Benton (2013). Why garden for wildlife? Social and ecological drivers, motivations and barriers for biodiversity management in residential landscapes. *Ecological Economics, 86*, 258–273.

Goulson, D. (2019). The insect apocalypse, and why it matters. *Current Biology, 29*(19), R967–R971.

Grade, A. M., Warren, P. S., & Lerman, S. B. (2022). Managing yards for mammals: Mammal species richness peaks in the suburbs. *Landscape and Urban Planning, 220*, 104337.

Gregory, R. L. (1997). Knowledge in perception and illusion. *Philosophical Transactions of the Royal Society of London. Series B: Biological Sciences, 352*(1358), 1121–1127.

Griffiths-Lee, J., Nicholls, E., & Goulson, D. (2022). Sown mini-meadows increase pollinator diversity in gardens. *Journal of Insect Conservation, 26*(2), 299–314.

Hall, D. M., & Martins, D. J. (2020). Human dimensions of insect pollinator conservation. *Current Opinion in Insect Science, 38,* 107–114.

Hamilton, A. M. (2005). Garden-centre valuations. *Journal of Retail & Leisure Property, 5*(1), 32–38.

Hand, K. L., Freeman, C., Seddon, P. J., Recioa, M. R., Stein, A., & van Heezik, Y. (2017). The importance of urban gardens in supporting children's biophilia. *PNAS, 114*(2), 274–279.

Heydon, J. (2020). Procedural environmental injustice in 'Europe's Greenest City': A case study into the felling of Sheffield's street trees. *Social Sciences, 9*(6), 100.

Hollow, M. (2011). Suburban ideals on England's interwar Council estates. *Journal of the Garden History Society, 39*(2), 203–217.

Huelskamp, A. C. (2018). Enhancing the health of school garden programs and youth: A systematic review. *Health Educator, 50*(1), 11–23.

Jones, R. E., Davis, K. L., & Bradford, J. (2012). The value of trees: Factors influencing homeowner support for protecting local urban trees. *Environment and Behavior, 45*(5) 650–676.

Knapp, J. L., Phillips, B. B., Clements, J., Shaw, R. F., & Osborne, J. L. (2021). Socio-psychological factors, beyond knowledge, predict people's engagement in pollinator conservation. *People and Nature, 3*(1), 204–220.

Laaksoharju, T., Rappe, E., & Kaivola, T. (2012). Garden affordances for social learning, play, and for building nature–child relationship. *Urban Forestry & Urban Greening, 11,* 195–203.

Langhamer, C. (2005). The meanings of home in postwar Britain. *Journal of Contemporary History, 40*(2), 341–362.

Lanner, J., Kratschmer, S., Petrović, B., Gaulhofer, F., Meimberg, H., & Pachinger, B. (2020). City dwelling wild bees: How communal gardens promote species richness. *Urban Ecosystems, 23*(2), 271–288.

LeBlanc, S. A., & Register, K. E. (2003). *Constant battles: Why we fight.* New York: St. Martins Press.

Lewis, S. L., & Maslin, M. A. (2015). Defining the anthropocene. *Nature, 519*(7542), 171–180.

Lindemann-Matthies, P., Mulyk, L., & Remmele, M. (2020). Garden plants for wild bees – Laypersons' assessment of their suitability and opinions on gardening approaches. *Urban Forestry & Urban Greening, 62,* 127181.

Lohr, V. I., & Pearson-Mims, C. H. (2005). Children's active and passive interactions with plants influence their attitudes and actions toward trees and gardening as adults. *HortTechnology, 15*(3), 472–476.

Meng-Lewis, Y., Xian, H., Lewis, G., & Zhao, Y. (2021). "Enthusiastic admiration is the first principle of knowledge and its last": A qualitative study of admiration for the famous. *SAGE Open, 11*(2), 21582440211006730.

Moore, D. (2014). My childhood was filled with secret places: The importance of secret places to children. *International Journal of Play, 3*(2), 103–106.

Mustapa, N. D., Malikia, N. Z., & Hamzah, A. (2018). Benefits of nature on children's developmental needs: A Review. *Asian Journal of Behavioural Studies, 3*(12), 31–42.

Onu, D., Kessler, T., & Smith, J. R. (2016). Admiration: A conceptual review. *Emotion Review, 8*(3), 218–230.

Orland, B., Vining, J., & Ebreo, A. (1992). The effect of street trees on perceived values of residential property. *Environment and Behavior, 24*(3), 298–325.

Rahimi, E., Barghjelveh, S., & Dong, P. (2022). A review of diversity of bees, the attractiveness of host plants and the effects of landscape variables on bees in urban gardens. *Agriculture & Food Security, 11*(1), 1–11.

Remmele, M., & Lindemann-Matthies, P. (2018). Like father, like son? On the relationship between parents' and children's familiarity with species and sources of knowledge about plants and animals. *Eurasia Journal of Mathematics, Science and Technology Education, 14*(10), em1581.

Richards, G., & Farrell, H. (2020). *Do bees need weeds?* London: Mitchell Beazley.

Shapiro, A. G., & Todorovic, D. (Eds.). (2016). *The Oxford compendium of visual illusions*. Oxford: Oxford University Press.

Stever, G. (2019). *The psychology of celebrity*. London: Routledge.

Sturm, U., Straka, T. M., Moormann, A., & Egerer, M. (2021). Fascination and joy: Emotions predict urban gardeners' pro-pollinator behaviour. *Insects, 12*(9), 785.

Taylor, N., Wright, J., & O'Flynn, G. (2021). Cultivating 'health' in the school garden. *Sport, Education and Society, 26*(4), 403–416.

van Heezik, Y. M., Dickinson, K. J. M., & Freeman, C. (2012). Closing the gap: Communicating to change gardening practices in support of native biodiversity in urban private gardens. *Ecology and Society, 17*(1), article 34.

Wake, S. J. (2008). In the best interests of the child: Juggling the geography of children's gardens (between adult agendas and children's needs). *Children's Geographies, 6*(4), 423–435.

Way, T. (2020). *Suburban gardens*. Stroud, Gloucestershire: Amberley Publishing.

Wilson, J. S., Forister, M. L., & Carril, O. M. (2017). Interest exceeds understanding in public support of bee conservation. *Frontiers in Ecology and the Environment, 15*(8), 460–466.

Woloszyn, M. R. (2010). Contrasting three popular explanations for the Muller-Lyer Illusion. *Current Research in Psychology, 1*(2), 102–107.

3
THE BEAUTY OF A GARDEN

There is a beauty, an aesthetic appeal, to a fine garden that rewards the observer's senses. What are the properties that distinguish a beautiful garden from the more mundane? There is a substantial philosophical literature that considers the appeal of the environment (Brady & Prior, 2020) of which gardens are a constituent part. Cooper (2006) notes that the admiration of a garden may rely on an appreciation of art alongside an appreciation of nature, although adding that it would be a mistake to force this overlap. It may be better, Cooper suggests, to consider art *and* nature so that the garden and its appreciation becomes an amalgam of the two, a specific manifestation of our engagement with nature.

In their review of visual aesthetics and human preference, Palmer et al. (2013) point to two indices of preference: (1) colour, either singly or in combination, along with colour harmony; and (2) spatial structure, including shape and composition. Palmer et al. also note that both American and British adults prefer cool colours (green, cyan and blue) to warm colours (red, orange and yellow). In terms of shape, Palmer et al. comment that "People tend to like objects with curved contours more than similar objects with sharp contours" (p. 92). These asthenic preferences for colour and shape may help in understanding how we perceive beauty in a garden.

Now, as beauty lies in the eye of the beholder, what may psychology contribute to understanding what we perceive to be beautiful? In accordance with the great garden designers discussed in Chapter 1 and in keeping with Palmer et al.'s review, the colour of the planting is a central element in garden design. Of course, as Berleant (2010) points out, there can never be universal agreement on what makes a scene beautiful to behold. This point is considered empirically by Karmanov and Hamel (2009), who asked two

DOI: 10.4324/9781003289661-4

groups of students, taking courses in either landscape architecture or psychology, to evaluate 12 different garden designs – designed by leading Dutch landscape architects, the 12 gardens at Makeblijde in Houten in the province of Utrecht form a garden complex, ranging in design from experimental to traditional – using a 22-item semantic differential scale incorporating items such as *ugly*, *impersonal* and *restful*. Karmanov and Hamel reported that, contrary to expectations, there was no difference in evaluations between the two groups. It may be that when it comes to gardens there is considerable uniformity in judging what is pleasing to the eye.

Colour in the Garden

We humans have a complex relationship with colour with preferences for some colours and a dislike of others; we form associations between mood and colour as in "feeling blue" or "seeing red"; we distinguish warm and cold colours; and there are psychological theories and therapies formed around colour preference. Of course, interest in colour goes beyond psychology and *colour theory* – understanding how colours alone and in combination create different visual effects and their influence on emotion – draws on disciplines such as art and design, physics, linguistics and philosophy (O'Connor, 2021).

At a physiological level, our perception of colour relies on two types of photoreceptor cells, *rods* and *cones*, in the eye's retina. These two types of cells are activated by light of different wavelengths: cones function best in relatively bright light and are responsible for colour vision, while rods work better in dim light and enable night vision. If the cones in an individual's eye do not function properly, this may cause colour blindness, most commonly where green and red are misperceived.

Psychological Effects of Colour

Colour Preference

Colour plays a significant role in many aspects of our everyday lives, influencing our choices in fashion, how we decorate where we live and work, our behaviour as consumers, our daily wellbeing, and even our levels of hostility towards others (e.g., Casas & Chinoperekweyi, 2019; Fetterman, Liu, & Robinson, 2015; Kodžoman, Hladnik, Čuden, & Čok, 2021; Smith, Metcalfe, & Lommerse, 2012; van der Voordt, Bakker, & de Boon, 2017). Once thought to be universal (e.g., Eysenck, 1941), it is now clear that there are individual differences in colour preference (Taylor, Clifford, & Franklin, 2013). However, there are social stereotypes that may shape colour preference as with "blue for boys and pink for girls" (LoBue & DeLoache, 2011); as for colour preference, these gender stereotypes are not universal (Davis et al., 2021).

Colour and Psychological Effect

Elliot (2019) charts psychological research dating from the 1800s, looking at cognitive variables such as time perception and attention in relation to colour perception. Another body of research has shown that colour can influence our emotional state (Gao & Xin, 2006; Gao et al., 2007; Valdez & Mehrabian, 1994). This latter work has defined the dimensions by which the effects of colour may vary: thus, there are hard and soft colours, masculine and feminine colours, and warm and cool colours. In addition, preferences may vary according to whether the judgement is being made for a single colour or a two-colour combination (Ou, Luo, Woodcock, & Wright, 2004a; Ou, Luo, Woodcock, & Wright, 2004b). Neale et al. (2021) looked at the effect of warm and cool-coloured gardens on psychological and physiological wellbeing. Two groups, all aged under 35 years, one American and the other British, viewed video recordings of static garden landscapes showing either warm colours or cool colours. The participants completed measures of subjective psychological wellbeing and biometric measures of stress, including heart rate and galvanic skin response. After viewing warm and cool gardens, the American participants showed an increase in pleasure along with a decrease in perceived stress. It appears that it is the effect of viewing a garden, rather than its colour composition, which has the beneficial effects. However, the positive pattern of responses to colour in the American sample was not evident in the British participants. It is unclear why the population difference appeared, leaving future research the task of explaining this variation. Nonetheless, Neale et al. provide support for the opinion that viewing a garden landscape may have a positive effect on our psychical and psychological functioning.

Colour and Personality

The idea that an individual's personality is related to their colour preference was developed by the Swiss psychotherapist Max Lüscher (1923–2017). In his 1948 book *Psychologie Der Farben* (Psychology of Colours), subsequently reprinted several times (Lüscher & Scott, 1969; Lüscher, 1990), Lüscher explains that the principle of the Colour Test is that hidden aspects of an individual's psychological functioning can be uncovered via their colour preferences. With its foundations formed through psychodynamic theory, the Lüscher Colour Test is a projective test through which an individual's unseen conflicts or fantasies can be revealed (Table 3.1).

As psychodynamic approaches fell away in mainstream psychology to be replaced by more scientific experimental methods, so the use of projective tests gave way to psychometrically robust personality tests. A Polish study by Wieloch et al. (2018) used the BFI-44 to look at personality and colour preference. The BFI-44 is a personality measure derived from the Five-Factor model

TABLE 3.1 Basic and Auxiliary Colours in the Lüscher Colour Test

Basic Colours

Blue Represents "Depth of Feeling" and is associated with tranquillity and contentment.
Green Represents "Elasticity of Will" and is associated with self-assertion and self-esteem.
Red Represents "Force of Will" and is associated with desire and domination.
Yellow Represents "Spontaneity" and is associated with originality and exhilaration.

Auxiliary Colours

Violet A mixture of red and blue, with properties of both, and is associated with reality in thought and desire, yet also enchantment and magic.
Brown Relates to bodily senses.
Black Expresses the idea of nothingness.
Grey Is neither dark or light and represents non-involvement and concealment.

of personality comprised of Extraversion, Agreeableness, Conscientiousness, Neuroticism and Openness to Experience (McCrae & John, 1992). Analysis of data from 144 respondents allowed Wieloch et al. to determine the significant associations between the five personality traits and colour preference. They found several associations, most notably that blue was a preference across all five traits, while Conscientiousness, Neuroticism and Openness were related to a preference for red.

In a Norwegian study by Watten and Fostervold (2021), 206 people (145 women and 61 men) completed the BFI-44 and nominated their favourite colour from six primary colours. Across the sample the rank order of the colour preferences was blue (42.2%), green (17%), red (13.6%), yellow (11.2%), black (8.3%) and white (7.8%). A strong preference for blue is also to be found in the Wieloch et al. study. When comparing men and women, Watten and Fostervold found no difference in preference for the chromatic colours (blue, green, red, yellow) although significantly more women than men preferred white and vice versa for black. There were significant differences on the personality measure according to preferred colour: when compared to the remainder of the sample those favouring blue had significantly higher scores on agreeableness and extraversion, while the red group gave lower scores on agreeableness. The blue group also had significantly higher scores on agreeableness and extraversion compared to the red group, and higher scores on agreeableness than the green group.

Given the varied psychological effects of colour, it follows that the introduction of colour through planting is an important consideration in landscape and garden design (Ender, Akdeniz, & Zencirkıran, 2016). It appears from the personality studies that some colours have a universal appeal, while others may appeal only to certain groups of people. However, the empirical evidence is limited, leaving a great deal more to be understood about the role of colour in the attraction of a garden.

Psychological Effects of Shape

Shape Preference

As with colour, do we prefer a particular shape? The weight of evidence suggests that we prefer to see a gentle, sweeping curve to sharp angular lines (Gómez-Puerto, Munar, & Nadal, 2016; Silvia & Barona, 2009). This preference for curved objects may be because unlike angular shapes, with their connotations of sharp and jagged edges, curves do not convey a sense of threat (Bar & Neta, 2006). Bertamini et al. (2016) conducted a series of empirical studies of shape preference and concluded that liking for curves and dislike of sharp lines may *not* be different sides of the same coin. They found that while there was a preference for curves, there was no evidence for an avoidance of angularity.

An awareness of a preference for curves is ingrained in expert designers. Bertamini and Sinico (2021) asked 56 expert designers, 31 females and 25 males, to draw various everyday objects of their own choice; they produced two versions, smooth and angular, of each object. The drawings were rated by 174 non-experts, 86 females and 88 males, on seven dimensions such as "ugly/beautiful", "old/modern" and "dangerous/safe". Bertamini and Sinico report that the ratings corroborated the association between smooth curves and perception of beauty.

The cognitive process involved in making judgements of preference may be affected by our level of emotional arousal. Leder, Tinio and Bar (2011) considered how the emotional valence, i.e., the perceived pleasantness or unpleasantness of a stimulus – such as a birthday cake or a razor blade – influences judgements of preference for curved objects. They found that observers expressed a preference for curved over sharp versions of the same stimulus, but only when the emotional valence of the stimulus was neutral or positive. The preference for curved stimuli was not found when the stimulus held a negative emotional valence. Thus, the degree of pleasantness we feel for a visual stimulus is constrained to non-threatening and positive objects.

Why we prefer curves – a preference for curves may be more pronounced in women rather than men (Palumbo, Rampone, & Bertamini, 2021) – is a matter for debate (Palmer, Schloss, & Sammartino, 2013). It may be that over time humans evolved an attraction to rounded forms, such as faces, which has become a generalised preference. Alternatively, the human visual system may be better at processing smooth rather than angular curvature (Bertamini, Palumbo, & Redies, 2019) so that there is an interplay between aesthetic preferences and our physiological and psychological processing of visual stimuli.

Psychological Effects of Scent

A garden's colour and shape engage our visual attention, while the scent from the plants attracts our olfactory attention. Of course, our senses do not function in isolation; for example, what we see visually is integrated with our olfactory perception to give an overall sensory experience (Gottfried, & Dolan, 2003). This point applies to the aesthetic experience of gardens as well as other types of environments (Tafalla, 2014). Indeed, the nature of the smell, say a particular plant or flower, may influence preference for particular landscapes (Zhao et al., 2018).

References

Bar, M., & Neta, M. (2006). Humans prefer curved visual objects. *Psychological Science, 17*(8), 645–648.

Berleant, A. (2010). Reconsidering scenic beauty. *Environmental Values, 19*(3), 335–350.

Bertamini, M., Palumbo, L., Gheorghes, T. N., & Galatsidas, M. (2016). Do observers like curvature or do they dislike angularity? *British Journal of Psychology, 107*(1), 154–178.

Bertamini, M., Palumbo, L., & Redies, C. (2019). An advantage for smooth compared with angular contours in the speed of processing shape. *Journal of Experimental Psychology: Human Perception and Performance, 45*(10), 1304–1318.

Bertamini, M., & Sinico, M. (2021). A study of objects with smooth or sharp features created as line drawings by individuals trained in design. *Empirical Studies of the Arts, 39*(1), 61–77.

Brady, E., & Prior, J. (2020). Environmental aesthetics: A synthetic review. *People and Nature, 2*(2), 254–266.

Casas, M., & Chinoperekweyi, J. (2019). Color psychology and its influence on consumer buying behavior: A case of apparel products. *Saudi Journal of Business and Management Studies, 4*, 441–456.

Cooper, D. E. (2006). *A philosophy of gardens*. Oxford: Oxford University Press.

Davis, J. T. M., Robertson, E., Lew-Levy, S., Neldner, K., Kapitany, R., & Nielsen, M. (2021). Cultural Components of sex differences in color preference. *Child Development, 92*(4), 1574–1589.

Elliot, A. J. (2019). A historically based review of empirical work on color and psychological functioning: Content, methods, and recommendations for future research. *Review of General Psychology, 23*(2) 177–200.

Ender, E., Akdeniz, N. S., & Zencirkıran, M. (2016). Colors and landscape. *Journal of Agricultural Faculty of Uludag University, 30* (Special Issue), 669–676.

Eysenck, H. J. (1941). A critical and experimental study of colour preferences. *The American Journal of Psychology, 54*(4), 385–394.

Fetterman, A. K., Liu, T., & Robinson, M. D. (2015). Extending color psychology to the personality realm: Interpersonal hostility varies by red preferences and perceptual biases. *Journal of Personality, 83*(1), 106–116.

Gao, X. P., & Xin, J. H. (2006). Investigation of human's emotional responses on colors. *Color Research & Application, 31*(5), 411–417.

Gao, X. P., Xin, J. H., Sato, T., Hansuebsai, A., Scalzo, M., Kajiwara, K., Guan, S., Valldeperas, J., Lis, M. J., & Billger, M. (2007). Analysis of cross-cultural color emotion. *Color Research & Application, 32*(3), 223–229.

Gómez-Puerto, G., Munar, E., & Nadal, M. (2016). Preference for curvature: A historical and conceptual framework. *Frontiers in Human Neuroscience*, 9, 712.

Gottfried, J. A., & Dolan, R. J. (2003). The nose smells what the eye sees: Crossmodal visual facilitation of human olfactory perception. *Neuron*, 39(2), 375–386.

Karmanov, D., & Hamel, R. (2009). Evaluations of design gardens by students of landscape architecture and non-design students: A comparative study. *Landscape Research*, 34(4), 457–479.

Kodžoman, D., Hladnik, A., Čuden, A. P., & Čok, V. (2021). Exploring color attractiveness and its relevance to fashion. *Color: Research and Application*, 47(1), 182–193.

Leder, H., Tinio, P. P., & Bar, M. (2011). Emotional valence modulates the preference for curved objects. *Perception*, 40(6), 649–655.

LoBue, V., & DeLoache, J. S. (2011). Pretty in pink: The early development of gender-stereotyped colour preferences. *British Journal of Developmental Psychology*, 29(3), 656–667.

Lüscher, M. (1948). *Psychologie Der Farben*. Basel: Test-Verlag.

Lüscher, M., & Scott, I. (1969). *The Lüscher color test*. New York: Random House.

Lüscher, M. (1990). *The Lüscher color test*. New York: Simon and Schuster.

McCrae, R. R., & John, O. P. (1992). An introduction to the five-factor model and its applications. *Journal of Personality*, 60(2), 175–215.

Neale, C., Griffiths, A., Chalmin-Pui, L. S., Mendu, S., Boukhechba, M., & Roe, J. (2021). Color aesthetics: A transatlantic comparison of psychological and physiological impacts of warm and cool colors in garden landscapes. *Wellbeing, Space and Society*, 2, 100038.

O'Connor, Z. (2021). Traditional colour theory: A review. *Color Research & Application*, 46(4), 838–847.

Ou, L. C., Luo, M. R., Woodcock, A., & Wright, A. (2004a). A study of colour emotion and colour preference. Part I: Colour emotions for single colours. *Color Research & Application*, 29(3), 232–240.

Ou, L. C., Luo, M. R., Woodcock, A., & Wright, A. (2004b). A study of colour emotion and colour preference. Part II: Colour emotions for two-colour combinations. *Color Research & Application*, 29(4), 292–298.

Palmer, S. E., Schloss, K. B., & Sammartino, J. (2013). Visual aesthetics and human preference. *Annual Review of Psychology*, 64, 77–107.

Palumbo, L., Rampone, G., & Bertamini, M. (2021). The role of gender and academic degree on preference for smooth curvature of abstract shapes. *PeerJ*, 9, e10877.

Silvia, P. J., & Barona, C. M. (2009). Do people prefer curved objects? Angularity, expertise, and aesthetic preference. *Empirical Studies of the Arts*, 27(1), 25–42.

Smith, D., Metcalfe, P., & Lommerse, M. (2012). Interior architecture as an agent for wellbeing. *Journal of the Home Economics Institute of Australia*, 19(3), 2–9.

Tafalla, M. (2014). Smell and anosmia in the aesthetic: Appreciation of gardens. *Contemporary Aesthetics (Journal Archive)*, 12(1), Article 19.

Taylor, C., Clifford, A., & Franklin, A. (2013). Color preferences are not universal. *Journal of Experimental Psychology: General*, 142(4), 1015–1027.

Valdez, P., & Mehrabian, A. (1994). Effects of color on emotions. *Journal of Experimental Psychology: General*, 123(4), 394–409.

van der Voordt, T., Bakker, I., & de Boon, J. (2017). Color preferences for four different types of spaces. *Facilities*, *35*(3/4), 155–169.

Watten, R. G., & Fostervold, K. I. (2021). Colour preferences and personality Traits. *Preprints*, 2021050642.

Wieloch, M., Kabzińska, K., Filipiak, D., & Filipowska, A. (2018). Profiling user colour preferences with BFI-44 personality traits. In W. Abramowicz & A. Paschke (Eds.), *International conference on business information systems* (pp. 63–76). Cham, Switzerland: Springer International Publishing.

Zhao, J., Huang, Y., Wu, H., & Lin, B. (2018). Olfactory effect on landscape preference. *Environmental Engineering & Management Journal*, *17*(6), 1483–1489.

4

GARDENS FOR THE PEOPLE

The suburban garden is typically tended by one or two gardeners, but there are several types of gardens where this is not the case. There are gardens created by and for groups of people, which serve a variety of functions, including horticulture, social interaction and education. There are also gardens held to have a therapeutic function; these are discussed in the following chapter.

Community Gardens

There are several types of community gardens, sometimes referred to under the umbrella term *urban agriculture* (e.g., Grebitus, 2021), with the common feature of being looked after by a group of people. The emphasis in community gardens is typically on environmental issues, the wellbeing and social inclusion of those who use the garden, and in many instances food production and distribution to the wider community. The gardening may be accompanied by other community events such as creative activities, playgroups, cooking and sharing food, and celebrations (Hou, 2017). Community gardens are found in many parts of the world from Australia (Kingsley, Foenander & Bailey, 2019), to Italy (Ruggeri, Mazzocchi & Corsi, 2016), to Sweden (Bonow & Normark, 2018) and beyond (Milbourne, 2021).

A great deal of effort is needed to establish a community garden: a location for the garden, suitable for the purpose, has to be found; oversight of the project established; there may be issues of accessibility; and guidance of unskilled enthusiasts (Tornaghi, 2014). Yet further, as illustrated by Pearson and Firth's (2012) survey of 47 community gardens in the East Midlands,

DOI: 10.4324/9781003289661-5

not all communal gardens have financial security. Pearson and Firth found that the majority of community gardens rely on external grants, with very few able to become financially independent.

The social psychological underpinning of the community garden lies in collective action; a group of volunteers organise themselves on an equal basis, perhaps united by a particular set of beliefs or principles (e.g., Braga Bizarria, Palomino-Schalscha & Stupples, 2022), to act collectively to acquire the benefits that the garden gives (Ioannou et al., 2016). It is axiomatic that an enterprise such as a community garden will need a system of governance, the nature of which will depend on the size and complexity of the project. Fox-Kämper et al. (2018) note several styles of governance, from top-down to bottom-up with various hybrids, all of which may change as the garden evolves over time.

A range of factors influences whether a project will become successful or struggle in the long-term (Doyle, 2022; Jacob & Rocha, 2021; Wesener et al., 2020). Fox-Kämper et al. describe the factors that enable a project to grow; these *enablers* include practicalities such as securing land tenure, engaging support from the local community and employing paid professionals when needed; alongside intangibles such as motivated volunteers and effective leadership. In contrast, the *barriers* to success include unsecured land tenure, an inability to engage with the community base and a lack of long-term government support.

Given the time and effort expended in establishing and maintaining a community garden, it might well be expected to deliver a range of benefits to its users. The benefits described by several researchers (Delshad, 2022; Feinberg et al., 2021; Poulsen et al., 2014; Ilieva et al., 2022; Quested et al., 2018) are classified and summarised in Table 4.1.

Although there are many benefits to be found in community gardening, it is not without its problems. It is highly unlikely that all community gardeners

TABLE 4.1 Benefits of Urban Community Gardening

Economic: Saving money on food, income generation through selling produce.
Educational: Sharing knowledge about gardens, mentoring new projects, improved gardening skills.
Environmental: Controlling the use of pesticides, improving brownfield city sites, increasing land accessibility, eco-friendly sustainability, sustaining green space.
Physical: Consuming fresh food, regular exercise, health benefits.
Psychological: Improved mood, relaxation, enjoyment of a spiritual connection to nature.
Social: Building social capital by creating connections and friendships between diverse people on the basis of cohesive civic engagement, community improvement and integration, promoting sustainable community development; mutual commitment to sharing work, seeds and plants.

will have the same motivations. In a study of seven Dutch community gardens, Veen et al. (2016) found that some gardeners were motivated by the social aspects, whereas others found growing vegetables more to their liking. If these differences between gardeners are accepted, then all is well; however, they can become a source of discontent. Any garden requires attention to maintain a desired standard; if the plot is large, as with many community gardens, then a considerable amount of time and work is required. If the effort expended outweighs the rewards, as may be the case if there are not enough gardeners, then individuals may decide to leave the project (Lee & Matarrita-Cascante, 2019).

The establishment of a community garden may accompany the process of gentrification of a part of a city. In this context the gentrification may have a *negative* effect, causing tensions both within the community and within the gardens. A study in California by Egerer and Fairbairn (2018) revealed how community gardens, some with exclusionary membership policies, become a mirror for wider tensions, such as racial disharmony and homelessness, and a target for theft and vandalism. In turn, these tensions spill over to conflicts between gardeners over issues such as installing security cameras and erecting fences. These tensions are, of course, not limited to California and may surface wherever urban agriculture is found (Di Fiore, Specht & Zanasi, 2021).

Allotments

The economic, political and social history of the allotment has generated considerable activity amongst scholars (Acton, 2011; Archer, 1997; Moselle, 1995). Many towns and cities contain plots of land that are available to rent for non-commercial gardening, providing the plot holder with a garden detached from their home. Way (2017) notes that "In 2015 there were 300,000 council-owned allotments in Britain ... . there were 100,000 people on allotment waiting lists" (p. 55); this figure includes allotments in England and Wales (Flavell, 2003), Ireland (Kettle, 2014) and Scotland (DeSilvey, 2003). Allotments are found across Europe (Poniży et al., 2021). Van den Berg et al. (2010) estimate that there are approximately 3,000,000 European allotments.

There is a tradition of the allotment as a crucial source of food during times of conflict; for example, in October 1939 the British Government launched the *Dig for Victory* campaign encouraging the nation to use their gardens and allotments to grow food to supplement rationing. The plea to *Dig for Victory* was endorsed by BBC Radio, which broadcast programmes such as *Back to the Land* and *The Radio Allotment,* while *The Kitchen Front* offered ideas on how to stretch further scanty supplies and improve their taste. The tradition of broadcasting about allotment gardening continues today on BBC Radio and via *The Allotment Podcast.*

It is evident that allotments remain a significant source of food production. Edmondson et al. (2020) looked at food production on allotment land provision in Leicester, United Kingdom. They estimated that for most crops the yield was comparable to that found in UK commercial horticulture, with allotments annually producing 1,200 tons of fruit and vegetables and 200 tons of potatoes. However, the allotment is not just about food production. Like a community garden, it can be a place for social gatherings or where families work together. Similarly, as with the suburban garden, the allotment can be a place for quiet individual reflection and contemplation. A study based in Texas by Lee and Matarrita-Cascante (2019) used web-based and on-site surveys to ask 180 gardeners, 119 females and 61 males, about the motivational factors that influenced their participation in allotment gardening. Lee and Matarrita-Cascante described three types of motivation: (1) functional such as growing food; (2) emotional as with feelings of enjoyment; and (3) conditional factors such as having the necessary time available.

The analysis revealed that an intricate mixture of motivations and socio-demographic characteristics, including age, predicted participation in allotment gardening. It was clear that older gardeners were motivated to produce fresh food, to improve their health and to be outdoors. The more an individual's self-concept was of themselves as a gardener, the greater their satisfaction with what the garden had to offer. As ever, time was of the essence: the less time claimed by the garden, the greater the chances of participation. However, for these particular gardeners, the personal outweighed the social: "The desire for social interaction with other gardeners was not found to be a significant functional motivator for participation in gardening" (p. 28).

Dobson et al. (2021) looked at the views of 96 allotment gardeners in England and Wales. These gardeners saw the primary purpose of their efforts as growing food with the additional benefits of feeling connected with nature, personal wellbeing and social interaction. The affinity for social contact is at odds with Lee and Matarrita-Cascante (2019), which may reflect sampling or methodological variations between the studies.

With some variation in emphasis, there is a clear overlap between community gardens and allotments. In the community garden, prominence is given to social connections between garden users and the local neighbourhood, with some attention given to food production. In the allotment, the position is reversed, with the individual gardener and their produce as the focus, with social issues assuming less importance.

School Gardens

The much-vaunted benefits of gardens have seen schools in several countries involve their pupils in gardening. There are claims that gardens in schools can have positive effects upon children's academic achievement, cognitive functioning, healthy eating, emotional development and prosocial attitudes and behaviour (Blair, 2009). This purported wide range of effects has generated a substantial literature on school gardens. There are obvious potential pitfalls in comparing research findings from different countries as well as comparing studies conducted in parts of the same country that have different educational systems for children of different ages. Nonetheless, the question remains: Are school gardens beneficial for the pupils?

Gardens and Academic Attainment

As Blair notes, school gardens have been used in the teaching of a wide range of both arts and science subjects. Williams and Dixon (2013) carried out a review of school gardens in the United States. They found that across different age groups and with a range of subjects, including arts, science, language, mathematics, social studies and writing, "The results of the studies show overwhelmingly that garden-based learning had a positive impact on students' grades, knowledge, attitudes, and behavior" (p. 225). These findings extend the positive benefits of a school garden beyond academic gains to other aspects of the pupils' behaviour. As academic performance is linked to cognitive functioning, a body of research has considered the effects of gardens and nature generally on cognitive functioning.

Vella-Brodrick and Gilowska (2022) reviewed 12 studies high on methodological quality on the effects of nature – including green playgrounds, outdoor learning, plants in classrooms, nature walks and scenes of nature from classroom windows – on school children's cognitive functioning. They concluded that the evidence is in favour of the proposition that exposure to nature enhances cognitive functioning in children and adolescents.

While Vella-Brodrick and Gilowska focused on nature providing a context for education, Mason et al. (2022) looked at the effects of short-term exposure to nature. They reviewed 14 studies, ranging from elementary school pupils to university students, where the exposure lasted between 10 and 90 minutes during the day. In 12 out of the 14 studies, there were benefits across all levels of educational provision in terms of attention and memory performance.

However, Williams and Dixon make two important points about the studies in their review. The first regards the quality of the research. The implementation of the strongest research designs, particularly in an emerging field of research, is a perennial problem in applied research (see Chapter 5 for a discussion of research designs). The use of weaker research designs reduces the degree of confidence in the findings.

The second point concerns research bias. Williams and Dixon make the comment that:

One of the major challenges in synthesizing research such as this is in locating bias, especially bias from those advocates passionate about the subject matter. Researcher bias is important to acknowledge. Those who both work and perform research in garden-based learning tend to be passionate advocates of the pedagogy, as was evident in many of the studies; yet the limitations that bias poses for research were not acknowledged. This field needs to begin to engage in deliberate, thoughtful, and critical analysis of their work (pp. 225–226).

As Williams and Dixon note, the possibility of research bias leading to inflated conclusions is increased in emerging fields of evaluation when the researcher is heavily invested, emotionally and ideologically, in the topic under investigation. The potential for research bias can be detected in several ways, such as the use of small unrepresentative samples and unstandardised measurement, not reporting null findings, and "cherry-picking" the data to produce favourable results.

Gardens and Pupils' Health

McCurdy et al. (2010) make the case that sedentary behaviour in childhood is linked with obesity and, in turn, a range of health issues such as asthma, Type 2 diabetes and vitamin D deficiency. They suggest that a focus on outdoor activities for children, including but not limited to gardening, and their families may assist in ameliorating the problems associated with obesity. There is some support for this proposition. Parmer et al. (2009) allocated 115 second-grade pupils (7–8 years of age) to one of three conditions: (1) nutrition education and gardening; (2) nutrition education only; or (3) control group. The pupils' knowledge about fruit and vegetables along with food preference and consumption was assessed before and after the programme. There was a positive outcome in that, compared to the control group, the pupils in the two nutrition groups showed significant improvements in knowledge, liking for and consumption of vegetables.

Qi et al. (2021) reported a systematic review and meta-analysis of 14 studies of the effects on childhood obesity of school gardening combined with physical activity. They found that gardening along with physical activity was associated with an increase in fruit and vegetable consumption, but there was no significant effect on two measures of health (body mass index and waist circumference).

Charlton et al. (2021) conducted a systematic review of school-based nutrition programmes, which included but was not limited to garden-based learning. The review included eleven garden-based studies: "All eleven of the gardening studies that were deemed as being unsuccessful, because of not achieving significant changes in the outcomes of interest (n 8) or having small sample sizes or lacking generalisability (n 3), were quasi-experimental and rated as either low (n 4) or very low (n 7) quality (p. 4655). These short-comings in design were not limited to garden-based programmes, and Charlton et al. comment on "The generally low quality of the cited studies" (p. 4658). There was little use of randomised designs studies, although protocols are available (Cosco et al., 2021), reflecting the view of Charlton et al. that there are practical and ethical issues in using randomisation in a school setting.

Chan, Tan and Gong (2022) conducted a systematic review of 35 studies addressing dietary and nutritional effects of school garden programmes. They concluded that the programmes can be effective in encouraging pupils' nutritional knowledge and attitudes towards vegetables and healthy eating. However, the effectiveness of a programme is likely to be moderated by a range of factors such as its length and whether there is any parental involvement.

Gardens and Pupils' Behaviour

Sakhvidi et al. (2022) carried out a review of the influence of long-term exposure to greenspaces, including but not limited to gardens, and childhood behaviour problems. The association between exposure to various types of green space and nine different behavioural outcomes – including attention-deficit/hyperactivity disorder, conduct problems, emotional symptoms, problems with peers and externalising and internalising disorders – was considered. Sakhvidi et al. reported that the majority of the associations were "suggestive" of a positive effect of green space exposure, but some negative associations were also evident. As have others, Sakhvidi et al. point to shortcomings in research design, such as a reliance on cross-sectional rather the longitudinal designs.

Putra et al. (2020) reported a systemic review of 15 studies of the asso-ciation between green space (including but not limited to gardens) and adolescent and childhood *prosocial* behaviours. They concluded their review with the statement that:

> The balance of evidence suggests that the development of prosocial behaviour may be associated with exposure to higher levels of nearby green space. However, the quality of this evidence is not yet sufficient to draw firm conclusions around causality or to offer specific guidance around well-defined interventions. Moreover, potential effect modifiers of the relationship between green space and prosocial behaviour were evident in some study contexts. Plausible mechanisms linking green space to prosociality have not been explored so far that need further investiga-tion. (p. 14)

Prison Gardens

There is a longstanding association between prisons and horticulture (Devine-Wright, Baybutt & Meek, 2019). In some prisons a farm, including livestock, was part of the estate and where prisoners worked to grow produce for the market, the funds raised going into the prison's coffers. Two further applications of prison horticulture are to enhance prisoner wellbeing and to assist in offender rehabilitation after release.

Prisoner Wellbeing

Baybutt and Chemlal (2016) make the case that a prison sentence affords the opportunity to change the behaviour of the criminal by improving their health and wellbeing, in turn lowering rates of reconviction. It is the case that many prisoners are from a socially disadvantaged background with a myriad of personal problems and are therefore in need of change. The argument is made that gardening may offer a vehicle for change to take place (DelSesto, in press; Jauk-Ajamie et al., 2023).

Horticultural Therapy

As Johnson (1999) notes, "Horticultural therapy is by no means a clear-cut practise with well-defined boundaries or methodology" (p. 225). It may be taken that, overall, horticultural therapy involves the use of plants and gardens towards a therapeutic or rehabilitative end with the aim of having a positive effect on the individual's wellbeing. This approach has been used to attempt to improve nutrition and health within prisoner populations (Thomas, 2022). A few small-scale studies have claimed psychological benefits following horticultural therapy in prisons (e.g., Lee et al., 2021; Rice & Remy, 1998; Timler, Brown & Varcoe, 2019). While the majority of prisoners are adult males, horticultural therapy has been used with a degree of success for both incarcerated juveniles (e.g., Twill, Purvis & Norris, 2011) and women (e.g., Jauk-Ajamie & Blackwood, in press).

An example of a prison-based horticultural programme is provided by Greener on the Outside for Prisons (GOOP) developed in 2008 for use in English prisons (Baybutt, Dooris & Farrier, 2019). GOOP is designed to address three aspects of prisoner functioning – healthy eating, mental wellbeing and physical activity – and runs as a collective programme between prisons, education and health providers, and a British university. Those prisoners who volunteer to participate in GOOP may take part in two activities: (i) those prisoners on temporary licence can partake in community-based projects such as landscaping; and (2) within the prison there may be opportunities to work on existing horticultural projects or to assist in the design of new gardens while taking part in accredited training.

Based on information gleaned from interviews and focus groups, Baybutt, Dooris and Farrier state that: "While limited in scale and scope ... . this study has revealed the profound and wide-ranging impacts attributed to participation in GOOP" (p. 978). It may be added that, of course, attributing an effect to a programme is not the same as empirically demonstrating that changes are actually brought about by the programme.

What happens to people held in prison is important, but for the majority of people it is what happens *after* a prison sentence that is of greater significance. A prison sentence serves several individual and social functions, one of which is to reduce the likelihood of reoffending. Given the time and cost invested in horticultural programmes in prison, it is surprising that so little is known about their effects on recidivism. Linden (2015) makes the point that evaluations of these prison programmes "Come mainly from internal programme evaluations based on small and self-selected samples that often have no clearly defined or matched reference group. Moreover, no systematic long-term evaluations of such programmes have been conducted" (p. 338). Given the limited evidence that horticultural community service programmes may be effective in reducing recidivism (Holmes & Waliczek, 2019), a full evaluation of the effect on recidivism of prison-based horticultural programmes is needed.

Immigrant Gardens

There are several reasons why people leave their homeland to live in a different country – some positive, others much less so. The desire to garden either individually or collectively in a style consistent with one's history, albeit in a different place and possibly climate, is understandable on the grounds of culture, spirituality, psychological solace, family, and plant preference, including growing edible produce (Biglin, 2020; Mazumdar & Mazumdar, 2012). Indeed, there are strong arguments that participating in communal gardens may facilitate immigrant health and cultural adjustment (Agustina & Beilin, 2012; Hartwig & Mason, 2016).

Guerrilla Gardening

It is tempting to think of gardeners as respectable, law-abiding types, mowing the grass and weeding the vegetables. However, recent years have seen a new type of gardener emerge, the so-called *Guerrilla Gardener*. The Spanish word *guerrilla* means "little war," referring to a type of warfare in which small groups, say armed citizens, employ tactics such as ambush against organised armed forces. The manual on how to conduct this style of combat was written by the legendary revolutionary Che Guevara (Guevara, 1961) (Figure 4.1).

FIGURE 4.1 Che Guevera.

Source: Photographer Emily Crawford.

The fusion of guerrilla tactics and gardening is described by Ralston (2012) as: "The political activity of reclaiming unused urban land, oftentimes illegally, for cultivation and beautification through gardening" (p. 57). In practise the guerrilla gardener takes land belonging to others such as roadside verges or neglected plots, typically without permission, and cultivates a garden. The use of people's property without their authorisation may be a criminal offence such as vandalism or trespass.

The practise of guerrilla gardening, which may have originated with the Green Guerrillas in Manhattan, mainly limited to the Global North and Australia (Hardman et al., 2018) can be found in several countries, including

Austria (Şolt & Heinz, 2017), Canada (Crane, Viswanathan & Whitelaw, 2013), Greece (Apostolopoulou & Kotsila, 2022), Kenya (Hussain, 2018) and the United States (McKay, 2011). As Certomà and Tornaghi (2015) note, there is clearly a political aspect to guerrilla gardening encompassing a range of political persuasions and subcultures:

> The claims expressed in the micro-politics of garden activism are quite diversified: DIY landscaping and engaged ecology, "digging for anarchy" and counter-neoliberal development, food sovereignty and the reconstruction of the urban commons, community empowerment and the "right to the city." (p. 1123)

Incidentally, gardening is not the only area of urban revolution: guerrilla knitting has become another revolutionary method of attempting to bring about political change (Greer, 2008).

Why Guerrilla Garden?

Why do some people feel the need to act, at times illegally, to create a green space, sometimes for growing vegetables, sometimes for aesthetic purposes? Millie (2023) interviewed six people, five females one male, who were members of various guerrilla gardening groups in the Northwest of England. While acknowledging that their actions were political, one interviewee stressed that this was not politics as aligned with a political party, which may have been off-putting to potential new recruits. The interviewees were aware that what they were doing was illegal but said that it was rare to run into opposition. They said that the police had donated materials, local authorities were interested and supportive, and local residents were appreciative of the changes to their neighbourhood. However, a UK study in the Midlands by Adams, Hardman and Larkham (2015) found less than universal acclaim for the activities of guerrilla gardeners. They conducted 59 interviews with people who lived close to a guerrilla garden. Alongside many positive comments, the interviewees made negative comments on the aloof behaviour of some of the gardeners, on being excluded from access to a once freely available site, and issues of ownership. As also suggested by Millie (2022), these issues of relationship and ownership are all familiar squabbles when social groups interact.

In summary, guerrilla gardening has become broadly accepted to the degree that it is a conventional form of breaking the law. As Millie (2022) puts it: "By challenging land ownership and an approved aesthetic order, the guerrilla gardener is a little bit rebellious – but this rebellion is broadly accepted by law enforcement and by wider culture (perhaps even endorsed by the BBC) and represents a normalised form of law-breaking" (p. 15). The final word goes to

Peter, one the guerrilla gardeners interviewed by Millie, who did not view his actions as rebellious, stating that "I don't feel like Che Guevara".

References

Acton, L. (2011). Allotment gardens: A reflection of history, heritage, community and self. *Papers from the Institute of Archaeology, 21*, 46–58.

Adams, D., Hardman, M., & Larkham, P. (2015). Exploring guerrilla gardening: Gauging public views on the grassroots activity. *Local Environment, 20*(10), 1231–1246.

Apostolopoulou, E., & Kotsila, P. (2022). Community gardening in Hellinikon as a resistance struggle against neoliberal urbanism: Spatial autogestion and the right to the city in post-crisis Athens, Greece. *Urban Geography, 43*(2), 293–319.

Archer, J. E. (1997). The nineteenth-century allotment: Half an acre and a row. *The Economic History Review, 50*(1), 21–36.

Agustina, I., & Beilin, R. (2012). Community gardens: Space for interactions and adaptations. *Procedia - Social and Behavioral Sciences, 36*, 439–448.

Baybutt, M., & Chemlal, K. (2016). Health-promoting prisons: Theory to practice. *Global Health Promotion, 23*(1_suppl), 66–74.

Baybutt, M., Dooris, M., & Farrier, A. (2019). Growing health in UK prison settings. *Health Promotion International, 34*(4), 792–802.

Biglin, J. (2020). Embodied and sensory experiences of therapeutic space: Refugee place-making within an urban allotment. *Health & Place, 62*, 102309.

Blair, D. (2009). The child in the garden: An evaluative review of the benefits of school gardening. *The Journal of Environmental Education, 40*(2), 15–38.

Bonow, M., & Normark, M. (2018). Community gardening in Stockholm: Participation, driving forces and the role of the municipality. *Renewable Agriculture and Food Systems, 33*(6), 503–517.

Braga Bizarria, M. T., Palomino-Schalscha, M., & Stupples, P. (2022). Community gardens as feminist spaces: A more-than-gendered approach to their transformative potential. *Geography Compass, 16*(2), e12608.

Certomà, C., & Tornaghi, C. (2015). Political gardening. Transforming cities and political agency. *Local Environment, 20*(10), 1123–1131.

Chan, C. L., Tan, P. Y., & Gong, Y. Y. (2022). Evaluating the impacts of school garden-based programmes on diet and nutrition-related knowledge, attitudes and practices among the school children: A systematic review. *BMC public Health, 22*(1), 1–33.

Charlton, K., Comerford, T., Deavin, N., & Walton, K. (2021). Characteristics of successful primary school-based experiential nutrition programmes: A systematic literature review. *Public Health Nutrition, 24*(14), 4642–4662.

Cosco, N. G., Wells, N. M., Monsur, M., Goodell, L. S., Zhang, D., Xu, T., ... & Moore, R. C. (2021). Research design, protocol, and participant characteristics of COLEAFS: A cluster randomized controlled trial of a childcare garden intervention. *International Journal of Environmental Research and Public Health, 18*(24), 13066.

Crane, A., Viswanathan, L., & Whitelaw, G. (2013). Sustainability through intervention: A case study of guerrilla gardening in Kingston, Ontario. *Local Environment, 18*(1), 71–90.

DelSesto, M. (in press). Therapeutic horticulture and desistance from crime. *The Howard Journal of Crime and Justice, 64*(4), 444–462.

Delshad, A. B. (2022). Community gardens: An investment in social cohesion, public health, economic sustainability, and the urban environment. *Urban Forestry & Urban Greening, 70,* 127549.

DeSilvey, C. (2003). Cultivated histories in a Scottish allotment garden. *Cultural Geographies, 10*(4), 442–468.

Devine-Wright, H., Baybutt, M., & Meek, R. (2019). Producing food in English and Welsh prisons. *Appetite, 143,* 104433.

Di Fiore, G., Specht, K., & Zanasi, C. (2021). Assessing motivations and perceptions of stakeholders in urban agriculture: A review and analytical framework. *International Journal of Urban Sustainable Development, 13*(2), 351–367.

Dobson, M. C., Reynolds, C., Warren, P. H., & Edmondson, J. L. (2021). "My little piece of the planet": The multiplicity of well-being benefits from allotment gardening. *British Food Journal, 123*(3), 1012–1023.

Doyle, G. (2022). In the garden: Capacities that contribute to community groups establishing community gardens. *International Journal of Urban Sustainable Development, 14*(1), 15–32.

Edmondson, J. L., Childs, D. Z., Dobson, M. C., Gaston, K. J., Warren, P. H., & Leake, J. R. (2020). Feeding a city – Leicester as a case study of the importance of allotments for horticultural production in the UK. *Science of the Total Environment, 705,* 135930.

Egerer, M., & Fairbairn, M. (2018). Gated gardens: Effects of urbanization on community formation and commons management in community gardens. *Geoforum, 96,* 61–69.

Feinberg, A., Hooijschuur, E., Rogge, N., Ghorbani, A., & Herder, P. (2021). Sustaining collective action in urban community gardens. *Journal of Artificial Societies and Social Simulation, 24*(3), 3.

Flavell, N. (2003). Urban allotment gardens in the eighteenth century: The case of Sheffield. *Agricultural History Review, 51*(1), 95–106.

Fox-Kämper, R., Wesener, A., Münderlein, D., Sondermann, M., McWilliam, W., & Kirk, N. (2018). Urban community gardens: An evaluation of governance approaches and related enablers and barriers at different development stages. *Landscape and Urban Planning, 170,* 59–68.

Grebitus, C. (2021). Small-scale urban agriculture: Drivers of growing produce at home and in community gardens in Detroit. *PLoS One, 16*(9), e0256913.

Greer, B. (2008). *Knitting for good! A guide to creating personal, social, and political change, stitch by stitch.* Boston, MA: Trumpeter Books.

Guevara, C. (1961). *Guerrilla warfare.* New York: Monthly Review Press.

Hardman, M., Chipungu, L., Magidimisha, H., Larkham, P. J., Scott, A. J., & Armitage, R. P. (2018). Guerrilla gardening and green activism: Rethinking the informal urban growing movement. *Landscape and Urban Planning, 170,* 6–14.

Hartwig, K. A., & Mason, M. (2016). Community gardens for refugee and immigrant communities as a means of health promotion. *Journal of Community Health, 41,* 1153–1159.

Holmes, M., & Waliczek, T. M. (2019). The effect of horticultural community service programs on recidivism. *HortTechnology, 29*(4), 490–495.

Hou, J. (2017). Urban community gardens as multimodal social spaces. In P. Y. Tan & C. Y. Jim (Eds.), *Greening cities: Forms and functions* (pp. 113–130). Singapore: Springer Nature.

Hussain, S. R. (2018). Reclaiming the city: Guerrilla gardening in Nairobi. In A. Catalani et al. (Eds.), *Cities' identity through architecture and arts* (pp. 389–395). London: Routledge.

Ilieva, R. T., Cohen, N., Israel, M., Specht, K., Fox-Kämper, R., Fargue-Lelièvre, A., ... & Blythe, C. (2022). The socio-cultural benefits of urban agriculture: A review of the literature. *Land, 11*(5), 622.

Ioannou, B., Morán, N., Sondermann, M., Certomà, C., Hardman, M., Anthopoulou, T., ... & Silvestri, G. (2016). Grassroots gardening movements: Towards cooperative forms of green urban development? In S. Bell et al. (Eds.), *Urban allotment gardens in Europe* (pp. 62–90). New York: Routledge.

Jacob, M., & Rocha, C. (2021). Models of governance in community gardening: Administrative support fosters project longevity. *Local Environment, 26*(5), 557–574.

Jauk-Ajamie, D., & Blackwood, A. (in press). *"I grow every day, like plants."* An evaluation of a gardening program for women in a residential community corrections setting. *Women & Criminal Justice*.

Jauk-Ajamie, D., Everhardt, S., Caruana, C. L., & Gill, B. (2023). Bourdieu in the women's prison garden: Findings from two clinical sociological garden interventions in the carceral field. *Journal of Applied Social Science, 17*(1), 92–110.

Johnson, W. T. (1999). Horticultural therapy: A bibliographic essay for today's health care practitioner. *Alternative Health Practitioner, 5*(3), 225–232.

Kettle, P. (2014). Motivations for investing in allotment gardening in Dublin: A sociological analysis. *Irish Journal of Sociology, 22*(2), 30–63.

Kingsley, J., Foenander, E., & Bailey, A. (2019). "You feel like you're part of something bigger": Exploring motivations for community garden participation in Melbourne, Australia. *BMC Public Health, 19*(1), 745.

Lee, A. Y., Kim, S. Y., Kwon, H. J., & Park, S. A. (2021). Horticultural therapy program for mental health of prisoners: Case report. *Integrative Medicine Research, 10*(2), 100495.

Lee, J. H., & Matarrita-Cascante, D. (2019). The influence of emotional and conditional motivations on gardeners' participation in community (allotment) gardens. *Urban Forestry & Urban Greening, 42*, 21–30.

Linden, S. (2015). Green prison programmes, recidivism and mental health: A primer. *Criminal Behaviour and Mental Health, 25*, 338.

McCurdy, L. E., Winterbottom, K. E., Mehta, S. S., & Roberts, J. R. (2010). Using nature and outdoor activity to improve children's health. *Current Problems in Pediatric and Adolescent Health Care, 40*(5), 102–117.

McKay, G. A. (2011). *Radical gardening: Politics, idealism & rebellion in the garden.* London: Frances Lincoln.

Mason, L., Ronconi, A., Scrimin, S., & Pazzaglia, F. (2022). Short-term exposure to nature and benefits for students' cognitive performance: A review. *Educational Psychology Review, 34*, 609–647.

Mazumdar, S., & Mazumdar, S. (2012). Immigrant home gardens: Places of religion, culture, ecology, and family. *Landscape and Urban Planning, 105*(3), 258–265.

Milbourne, P. (2021). Growing public spaces in the city: Community gardening and the making of new urban environments of publicness. *Urban Studies, 58*(14), 2901–2919.

Millie, A. (2023). Guerrilla gardening as normalised law-breaking: Challenges to land ownership and aesthetic order. *Crime, Media, Culture, 19*(2), 191–208.

Moselle B. (1995). Allotments, enclosure, and proletarianization in early nineteenth-century Southern England. *Economic History Review, 48*(3), 482–500.

Parmer, S. M., Salisbury-Glennon, J., Shannon, D., & Struempler, B. (2009). School gardens: An experiential learning approach for a nutrition education program to increase fruit and vegetable knowledge, preference, and consumption among second-grade students. *Journal of Nutrition Education and Behavior, 41*(3), 212–217.

Pearson, D. H., & Firth, C. (2012). Diversity in community gardens: Evidence from one region in the United Kingdom. *Biological Agriculture & Horticulture, 28*(3), 147–155.

Poniży, L., Latkowska, M. J., Breuste, J., Hursthouse, A., Joimel, S., Külvik, M., Leitão, T. E., ... & Kacprzak, E. (2021). The rich diversity of urban allotment gardens in Europe: Contemporary trends in the context of historical, socioeconomic and legal conditions. *Sustainability, 13*, 11076.

Poulsen, M. N., Hulland, K. R., Gulas, C. A., Pham, H., Dalglish, S. L., Wilkinson, R. K., & Winch, P. J. (2014). Growing an urban Oasis: A qualitative study of the perceived benefits of community gardening in Baltimore, Maryland. *Culture, Agriculture, Food and Environment, 36*(2), 69–82.

Putra, I. G. N. E., Astell-Burt, T., Cliff, D. P., Vella, S. A., John, E. E., & Feng, X. (2020). The relationship between green space and prosocial behaviour among children and adolescents: A systematic review. *Frontiers in Psychology, 11*, 859.

Qi, Y., Hamzah, S. H., Gu, E., Wang, H., Xi, Y., Sun, M., ... & Lin, Q. (2021). Is school gardening combined with physical activity intervention effective for improving childhood obesity? A systematic review and meta-analysis. *Nutrients, 13*(8), 2605.

Quested, E., Thøgersen-Ntoumani, C., Uren, H., Hardcastle, S. J., & Ryan, R. M. (2018). Community gardening: Basic psychological needs as mechanisms to enhance individual and community well-being. *Ecopsychology, 10*(3), 173–180.

Ralston, (2012). A Deweyan defense of guerrilla gardening. *The Pluralist, 7*(3), 57–70.

Rice, J. S., & Remy, L. L. (1998). Impact of horticultural therapy on psychosocial functioning among urban jail inmates. *Journal of Offender Rehabilitation, 26*(3-4), 169–191.

Ruggeri, G., Mazzocchi, C., & Corsi, S. (2016). Urban gardeners' motivations in a metropolitan city: The case of Milan. *Sustainability, 8*, 1099.

Sakhvidi, M. J. Z., Knobel, P., Bauwelinck, M., de Keijzer, C., Boll, L. M., Spano, G., ... & Dadvand, P. (2022). Greenspace exposure and children behavior: A systematic review. *Science of the Total Environment*, 153608.

Şolt, B. H., & Heinz, G. K. (2017). Urban gardens in Vienna and Istanbul: A review. *İdealkent, 8*(21), 159–180.

Thomas, A. (2022). Developing an evidence-based nutrition curriculum for correctional settings. *Journal of Correctional Health Care, 28*(2), 177- 128.

Timler, K., Brown, H., & Varcoe, C. (2019). Growing connection beyond prison walls: How a prison garden fosters rehabilitation and healing for incarcerated men. *Journal of Offender Rehabilitation, 58*(5), 444–463.

Tornaghi, C. (2014). *How to set up your own urban agricultural project with a socio-environmental justice perspective. A guide for citizens, community groups and third sector organisations*. Leeds: The University of Leeds.

Twill, S., Purvis, T., & Norris, M. (2011). Weeds and seeds: Reflections from a gardening project for juvenile offenders. *Journal of Therapeutic Horticulture*, *21*(1), 6–17.

Van den Berg, A., Van Winsum-Westra, M., de Vries, S., & Van Dillen, S. (2010). Allotment gardening and health: A comparative survey among allotment gardeners and their neighbours without an allotment. *Environmental Health*, *9*, Article Number 74.

Veen, E. J., Bock, B. B., Van den Berg, W., Visser, A. J., & Wiskerke, J. S. (2016). Community gardening and social cohesion: Different designs, different motivations. *Local Environment*, *21*(10), 1271–1287.

Vella-Brodrick, D. A., & Gilowska, K. (2022). Effects of nature (greenspace) on cognitive functioning in school children and adolescents: A systematic review. *Educational Psychology Review*, *34*(3), 1217–1254.

Way, T. (2017). *Allotments*. Stroud, Gloucestershire: Amberley Publishing.

Wesener, A., Fox-Kämper, R., Sondermann, M., & Münderlein, D. (2020). Placemaking in action: Factors that support or obstruct the development of urban community gardens. *Sustainability*, *12*(2), 657.

Williams, D. R., & Dixon, P. S. (2013). Impact of garden-based learning on academic outcomes in schools: Synthesis of research between 1990 and 2010. *Review of Educational Research*, *83*(2), 211–235.

5

THE PSYCHOLOGY OF THE GARDENER

As time unfolded, grand expansive gardens, once the province of the rich and powerful, became an ordinary, everyday adjunct to thousands of town and country dwellings. This sociological change precipitated an exponential growth in the number of gardeners, which, in turn, led to more social change to meet the needs of this new market. There was an appetite for the ever more sophisticated radio and television programmes and the celebrities they created. As discussed in Chapter 2, the new commercial opportunities saw the rise of garden centres, later enhanced by online retailers, giving the gardener an ever-widening range of choice not just in plant varieties but also in a range of gardening paraphernalia.

The new wave of post-war gardening assumed diverse forms, including community gardening and guerrilla gardening, but at the heart of the matter there remains the individual gardener. Thus, the question considered in this chapter is why do gardeners garden? Why do some people spend inordinate amounts of their time and money digging and planting and sometimes worrying about their plot? To try to formulate an answer to this question, consideration is given to what drives the individual's behaviour. An obvious answer to this question is that people garden because they find it rewarding. The relationship between an individual's behaviour and the rewards it produces is at the heart of *learning theory*.

Learning Theory

How people acquire behaviour was the life's work of the great American psychologist B. F. Skinner (1904–1990). Skinner formulated the principles of reinforcement to account for the acquisition and maintenance of behaviour.

DOI: 10.4324/9781003289661-6

Principles of Reinforcement

Skinner's learning theory, called *operant learning*, makes the distinction between two contingencies – *positive reinforcement* and *negative reinforcement* (Skinner, 1974). With *positive* reinforcement the individual *gains* the rewarding consequences resulting from the effects of their behaviour on the environment. Thus, when the occasion arises, he or she is likely to repeat that behaviour to gain these rewards. *Negative* reinforcement (entirely different from *punishment*, which is when a behaviour *decreases* in frequency) takes place when the consequences of that behaviour avoid an unwanted outcome. Again, when the occasion arises the individual will act to circumvent the unwanted outcome and therefore is being negatively reinforced. Thus, a behaviour that is repeated over time is being reinforced. Nye (1992) offers a good account of the difference between the two reinforcement contingencies.

To give a relevant example: the behaviour of the gardener whose blooms win prises at the flower show, whose efforts attract admiring comments from friends, who holds social events in their garden, and who gardens with their partner (although this can cut one of two ways!), is *positively* reinforced by the rewards gained by their behaviour. The gardener who is satisfied that their garden does not look a mess and avoids comments from unhappy neighbours is *negatively* reinforced. Of course, a single behaviour can produce a range of consequences, positive and negative, but the definitional point remains: a behaviour that is repeated over time is being reinforced by the external rewards it produces.

Contemporary learning theory, termed *social learning theory* (Bandura, 1977), also holds that, as in operant learning, a behaviour's rewarding consequences may be external to the person. However, social learning theory adds a new class of rewards: the thoughts and feelings that are *internal* to the individual. Thus, following this approach a gardener may be rewarded by their feelings of achievement, creativity, enjoyment and personal satisfaction.

Not all gardeners experience the same rewards; nor do the same rewards appeal to all gardeners. Thus, some gardeners prize creative garden design over plants; others find pleasure in collecting varieties or hybrids of the same plant, even contributing to national collections (see below). Some gardeners enjoy quiet individual contemplation; others compete at garden shows. Interviews with active gardeners have revealed that a range of rewards is presented by being in the garden. Bhatti et al. (2009) drew on written records from the Mass Observation Archive at the University of Sussex. This archive is essentially a writing project in which a panel of men and women write about their lives, some of which include gardening. Bhatti et al. discerned three aspects of everyday gardening: an awareness of sensory perception, particularly touch, of their environment; cultivation of the garden and caring for oneself and others; and emotional attachments relating to personal memories. The universality of these experiences is reflected in a study conducted in Oman where gardeners reported similar views about their relationship with gardening (Al-Mayahi et al., 2019).

If we translate these reported experiences into the rewards gardening provides then we may make the distinction between personal and social rewards.

Personal Rewards

Physical Rewards

Like walking the dog, the physical activity associated with gardening brings some health benefits. A United Kingdom study reported by Chalmin-Pui et al. (2021a) collected data from 5,766 gardeners and found that those who gardened at least two or three times a week perceived health benefits in terms of lower stress and maintaining their general wellbeing. However, the gardeners said that the health benefits took second place to the pleasures of gardening, such as watching the plants grow in a well-maintained space. Gross and Lane (2007) make the point that the rewards of gardening change over the lifespan. They conducted interviews with female and male gardeners between 18 and 85 years old whose "Accounts clearly illustrate the way in which the garden and the act of gardening reflect concerns, interests, aspirations and emotions throughout the lifespan, acting as a location for coping with stresses, creating identities or dealing with loss" (p. 239). At the older end of the age range, Scott et al. (2015) conducted a survey of 331 Australian gardeners aged between 60 and 90 years. They found that the older gardeners saw physical activity as both beneficial in itself and helpful for ailments such as arthritis. The elderly gardeners said that as they grew older, they had to modify their gardening techniques.

Psychological Rewards

Gardeners can feel very close to their gardens, as found by Freeman et al. (2012) from interviews with 55 householders in Dunedin, New Zealand. The interviewees said that their gardens were important for several reasons: (i) to improve physical and psychological health; (ii) to express ownership and identity; (iii) to serve as a place for socialisation; (iv) to connect with nature; and (v) to grow domestic produce. Scott et al. (2015) reported that their Australian gardeners said that gardening had a restorative effect on their emotions, instilling feelings of peace and relieving stress. This restorative effect, Scott et al. suggest, is in keeping with biophilia theory and our need to have contact with nature (see Chapter 2).

Collecting

People, mainly men (Apostolou, 2011), like to collect things; around the world there are vast collections of objects on public display in museums and art galleries. There are large private collections of valuables such as antiques, art, classic cars, fine wines and rare books, as well as countless small-scale, sometimes valuable, collections of a myriad of objects, including those noted above as well as autographs, comic books, coins, postage stamps, vinyl records and the list goes on. Indeed, it seems almost anything is collectable, ranging from beer mats and paperweights to seashells and luggage. Unlike collections in public institutions, many private collections remain hidden from general view.

Why do we give so much time and money to collecting? Following their systematic review, Lee, Brennan and Wyllie (2022) suggest that the collector's behaviour is underpinned by one or more motivations. The collector may gain a *sense of achievement* as they partially or fully complete their collecting goals; there may be *social rewards* through communicating with and meeting fellow collectors; collectors may *cooperate* or *compete* in pursuit of an item; collecting can prompt both *societal memories*, as seen with sporting memorabilia and vintage clothes, while objects such as books and toys evoke *personal memories*; a collection provides a *legacy,* by which the collector can be remembered and the objects passed on; and, last but not least, the collection may accrue a *financial value* and serve as an investment for the future. There are other considerations: Apostolou (2011) suggests that the value of a collectible item is a function of its aesthetic qualities and rarity, and for some objects physical size is important. Lee, Brennan and Wyllie's (2022) suggestion that the study of collecting lacks a coherent theoretical framework is reflected in the plethora of theories of collecting from a range of disciplines, including biology, evolution, personality theory, Freudian theory and social psychology (Apostolou, 2011; Delbourgo, 2019; Kleine, Peschke, & Wagner, 2021; McIntosh & Schmeichel, 2004).

Some gardeners collect a certain type of plant, say dahlias or roses, while those gardeners with the necessary resources take care of a part of the National Plant Collection. The National Plant Collection is a documented collection of groups of plants generally linked botanically by plant group. In the United Kingdom, there are about 95,000 plants, held across 650 collections based in gardens, greenhouses, allotments and indoors; the collections are cared for by individual gardeners, in botanic gardens and plant nurseries, and in Local Authority parks. The collections are a significant resource for all gardeners, growers and researchers concerned with plant conservation, ensuring they will be preserved for future generations. The collections are made available to the public by appointment or on open days.

Competitions

There can be no doubt that competition is a fact of life, and if there are prizes, then so much the better. Competitive gardeners may exhibit their flowers and vegetables in various categories – such as dahlias or tulips, apples or leeks – at a range of shows. Thus, gardeners at an allotment may hold their own local competition, geographical areas may have their own show, such as the Harrogate Flower Show, and there are shows on a national, even international, scale such as the Royal Horticultural Society Chelsea Flower Show (called the "world cup of gardening"). Gardeners will go to extraordinary lengths to win a show; for example, UK records include the longest leek at 1.432 metres and the heaviest field pumpkin weighing in at 121.6 kilograms.

The Lawn

The garden gives ample opportunity to interact with nature while moulding it to one's wishes. The lawn is a feature of many British gardens where it is, sometimes obsessively, mown, fed and nurtured. The lawn may have been created in Versailles where it was called *tapis vert* (small carpet), and it became popular in those countries, including Britain, where it easily adapted to a temperate climate. Many people see a lawn as an essential feature of gardens and parks, but it comes at a price. A good lawn demands high levels of maintenance and considerable amounts of water, fertiliser and weed killer, all with their attendant impact on the pocket and to the detriment of other plants (Watson et al., 2020). To save time, money and chemical pollutants, alternatives to the traditional lawn are appearing, cultivated by the so-called "lawn dissidents" (Lebowitz & Trudeau, 2017). The alternative choice of an artificial lawn is becoming more prevalent, as seen by the growing number of commercial outlets for artificial grass (Brooks & Francis, 2019) and their increasing popularity (e.g., Henderson, Perkins, & Nelischer, 1998). There are diametrically opposing views on artificial lawns. Some people are strongly opposed on aesthetic and environmental grounds; others see them as a boon because they look good with a minimum of effort (Barnes & Watkins, 2022; Brooks & Francis, 2019; Harris et al., 2013) (Figure 5.1).

As the effects of climate change become more pronounced, it may be that lawns become an endangered aspect of the traditional garden. However, with a careful management regime grass can be resilient to fluctuations in heat and rainfall (Trudgill, Jeffery, & Parker, 2010). Yet further, artificial lawns are not as eco-friendly as may be thought as the rubber compounds from which they are made may decompose over time. The products of this decomposition, such as volatile organic compounds and polycyclic aromatic hydrocarbons, may be hazardous to the environment and to health (Cheng, Hu, & Reinhard, 2014; Francis, 1995).

FIGURE 5.1 Contemplating the Garden.

Source: Photographer Anthony Wade.

Source: Unsplash

Despite the controversies, there is no doubt that an artificial lawn can look good. A little while ago I was wandering around an open garden event at a local village. I walked into a smallish garden with neat and tidy flower beds set around the centrepiece of an immaculate green rectangle. "Your lawn looks really good". I said to the owner. "Yes", she replied, "I gave it a good going over with the vacuum cleaner this morning".

Creativity

As outlined by Funke (2009), the study of creativity is of longstanding interest in psychological research. The garden provides a perfect environment for the creative person to display their use of colour, choice of design and selection of ornaments. The creative dimension of structure used to form the garden speaks to an essential part of our psychological functioning. In a complex world, we use cognitive structures to order our world and make life easier and more efficient. There are individual differences in the need for structure (Neuberg & Newsom, 1993), and Van den Berg and van Winsum-Westra (2010) suggest that: "Three basic types of gardens can be distinguished, which represent a continuum ranging from formal to informal garden styles: manicured, romantic, and wild" (p. 180). A manicured garden, as the name suggests, is a formal garden with neat lines and orderly planting. A romantic garden has some of the attributes of a manicured garden but with more profuse, lush planting. Finally, the wild garden is the opposite of a manicured garden, informal and unstructured with plants allowed to grow as and where they will.

An individual's need for structure influences their preferred style of garden. As may be predicted, Van den Berg and van Winsum-Westra found that gardeners with a high need for structure, as compared to those with a low need for structure, were more likely to own a manicured or romantic garden and less likely to own a wild garden. In addition, men were almost three times more likely to own a manicured rather than a wild garden; the likelihood of owning a romantic rather than a manicured garden fell with advancing age.

Driven by climate change there has been a growth in *climate-adapted gardens,* which are designed to be resilient to climate change. This type of garden uses fewer native plants in favour of exotic plants better suited to high temperatures and lower rainfall. Hoyle (2021) reported that people gave favourable reactions to climate-adapted gardens in terms of their aesthetic appeal when contrasted with other styles.

Coping with Loss

Loss comes in different forms: we may lose a loved one, leave our place of birth, or lose a personal faculty. Our memories, emotions and sense of loss may be evoked by cues in our everyday environment, which for some includes the garden where many hours have been spent (Ginn, 2014). Gardens may hold cues that trigger memories of loss such as a particular walk through the garden, a garden tool belonging to a father, or a plant introduced into the garden to mark a birthday. We may add features to the garden that serve to commemorate our loss – a memorial flower bed, a statement plant, the ubiquitous memorial bench – but which may accentuate absence. Photographs of long-gone gardens from childhood may arouse similar memories and feelings.

Those people who leave their homeland to make a life in a new country lose many familiar environmental features (see Chapter 4). Mazumdar and

Mazumdar (2012) conducted interviews with 28 immigrants, responsible for 16 home gardens, who had moved to Southern California from China, India, Indonesia, Iran, the Philippines, Taiwan and Vietnam. Mazumdar and Mazumdar suggest that "Immigrants appropriate their backyards to create distinctive culture spaces" (p. 264): this is manifest in the garden's use as a place for cultural purposes such as to honour and commemorate family members, grow the plants necessary for their culture's cuisine, and facilitate religious practises.

Happiness

Happiness means different things to different people while having the common denominators of positive thoughts and feelings connected to certain physical and social settings. An American study by Ambrose et al. (2020) found that for personal happiness, respondents ranked gardening alongside activities such as cycling and walking and above others such as walking and shopping. Growing vegetables was rated as more pleasurable than ornamental gardening, while gardening alone is no more or less pleasurable than gardening with company. As noted above, the gardeners in the Bhatti et al. study spoke of echoes of happiness from childhood experiences in the garden.

Social Rewards

Friends and Family

The garden is a setting for socialising with friends and family and for children to play alone or with friends. As discussed in Chapter 2, a garden either at home or at school can play a role in a child's development, particularly in fostering their connection with nature. The influence of the garden in the child's later life is evident in adult memories of time in the garden. Francis (1995) interviewed over 100 adult gardeners from California and Norway, recording their memories of childhoods spent in gardens. The common elements of these memories concerned particular flowering plants, vegetables and trees; water; play structures; and places to play and to hide. These interactions with gardens contrast with some aspects of modern childhood. Francis remarks that: "One of the top ten selling computer games for children is 'The Backyard', a computer simulation of playing outside. These trends have led me to characterise childhood today as The Childhood of Imprisonment" (p. 188).

If a garden at the back of the house is a private place, then quite the reverse is true of the suburban front garden. Chalmin-Pui et al. (2021b) conducted focus groups for gardeners taking part in *Britain in Bloom*, a national campaign organised by the UK Royal Horticultural Society for people to enhance their locality through gardening. With regard to front gardens, there were four main themes. First, *self-identity*, whereby gardeners see their front gardens and their gardening as a significant aspect of their personal identity and self-expression. Second, *community* as their efforts add to the local area, giving pleasure to neighbours and those passing by. Third, *fulfilment* in the pleasure of achievement and pleasure in all that the garden gives. Fourth, *health benefits* from physical activity and the social and psychological rewards of community and fulfilment.

References

Al-Mayahi, A., Al-Ismaily, S., Gibreel, T., Kacimov, A., & Al-Maktoumi, A. (2019). Home gardening in Muscat, Oman: Gardeners' practices, perceptions and motivations. *Urban Forestry & Urban Greening*, *38*, 286–294.

Ambrose, G., Das, K., Fan, Y., & Ramaswami, A. (2020). Is gardening associated with greater happiness of urban residents? A multi-activity, dynamic assessment in the Twin-Cities region, USA. *Landscape and Urban Planning*, *198*, 103776.

Apostolou, M. (2011). Why men collect things? A case study of fossilised dinosaur eggs. *Journal of Economic Psychology*, *32*(3), 410–417.

Bandura, A. (1977). *Social learning theory*. Englewood Cliffs, NJ: Prentice Hall.

Barnes, M. R., & Watkins, E. (2022). Differences in likelihood of use between artificial and natural turfgrass lawns. *Journal of Outdoor Recreation and Tourism*, *37*, 100480.

Bhatti, M., Church, A., Claremont, A., & Stenner, P. (2009). 'I love being in the garden': Enchanting encounters in everyday life. *Social & Cultural Geography*, *10*(1), 61–76.

Brooks, A., & Francis, R. A. (2019). Artificial lawn people. *Environment and Planning E: Nature and Space*, *2*(3), 548–564.

Chalmin-Pui, L. S., Roe, J., Griffiths, A., Smyth, N., Heaton, T., Clayden, A., & Cameron, R. (2021a). "It made me feel brighter in myself" – The health and well-being impacts of a residential front garden horticultural intervention. *Landscape and Urban Planning*, *205*, 103958.

Chalmin-Pui, L. S., Griffiths, A., Roe, J., & Cameron, R. (2021b). Gardens with kerb appeal – A framework to understand the relationship between Britain in Bloom gardeners and their front gardens. *Leisure Sciences*, 1–21.

Cheng, H., Hu, Y., & Reinhard, M. (2014). Environmental and health impacts of artificial turf: A review. *Environmental Science & Technology*, *48*(4), 2114–2129.

Delbourgo, J. (2019). Collect or die. *British Journal for the History of Science: Themes*, *4*, 273–281.

Francis, M. (1995). Childhood's garden: Memory and meaning of gardens. *Children's Environments*, *12*(2), 183–191.

Freeman, C., Dickinson, K. J., Porter, S., & Van Heezik, Y. (2012). "My garden is an expression of me": Exploring householders' relationships with their gardens. *Journal of Environmental Psychology*, *32*(2), 135–143.

Funke, J. (2009). On the psychology of creativity. In P. Meusburger, J. Funke, & E. Wunder (Eds.), *Milieus of creativity: An interdisciplinary approach to spatiality of creativity* (pp. 11–23). Dordrecht: Springer.

Ginn, F. (2014). Death, absence and afterlife in the garden. *Cultural Geographies, 21*(2), 229–245.

Gross, H., & Lane, N. (2007). Landscapes of the lifespan: Exploring accounts of own gardens and gardening. *Journal of Environmental Psychology, 27*(3), 225–241.

Harris, E. M., Martin, D. G., Polsky, C., Denhardt, L., & Nehring, A. (2013). Beyond "Lawn People": The role of emotions in suburban yard management practices. *The Professional Geographer, 65*(2), 345–361.

Henderson, S. P., Perkins, N. H., & Nelischer, M. (1998). Residential lawn alternatives: A study of their distribution, form and structure. *Landscape and Urban Planning, 42*(2-4), 135–145.

Hoyle, H. E. (2021). Climate-adapted, traditional or cottage-garden planting? Public perceptions, values and socio-cultural drivers in a designed garden setting. *Urban Forestry & Urban Greening, 65*, 127362.

Kleine, J., Peschke, T., & Wagner, N. (2021). Collectors: Personality between consumption and investment. *Journal of Behavioral and Experimental Finance, 32*, 100566.

Lee, C., Brennan, S., & Wyllie, J. (2022). Consumer collecting behaviour: A systematic review and future research agenda. *International Journal of Consumer Studies, 46*(5), 2020–2040.

Lebowitz, A., & Trudeau, D. (2017). Digging in: Lawn dissidents, performing sustainability, and landscapes of privilege. *Social & Cultural Geography, 18*(5), 706–731.

McIntosh, W. D., & Schmeichel, B. (2004). Collectors and collecting: A social psychological perspective. *Leisure Sciences, 26*(1), 85–97.

Mazumdar, S., & Mazumdar, S. (2012). Immigrant home gardens: Places of religion, culture, ecology, and family. *Landscape and Urban Planning, 105*(3), 258–265.

Neuberg, S. L., & Newsom, J. T. (1993). Personal need for structure: Individual differences in the desire for simpler structure. *Journal of Personality and Social Psychology, 65*(1), 113–131.

Nye, R. D. (1992). *The legacy of B. F. Skinner: Concepts and perspectives, controversies and misunderstandings.* Pacific Grove, CA: Brooks/Cole Publishing Company.

Scott, T. L., Masser, B. M., & Pachana, N. A. (2015). Exploring the health and wellbeing benefits of gardening for older adults. *Ageing & Society, 35*(10), 2176–2200.

Skinner, B. F. (1974). *About behaviourism.* London: Jonathon Cape.

Trudgill, S., Jeffery, A., & Parker, J. (2010). Climate change and the resilience of the domestic lawn. *Applied Geography, 30*(1), 177–190.

Van den Berg, A. E., & van Winsum-Westra, M. (2010). Manicured, romantic, or wild? The relation between need for structure and preferences for garden styles. *Urban Forestry & Urban Greening, 9*(3), 179–186.

Watson, C. J., Carignan-Guillemette, L., Turcotte, C., Maire, V., & Proulx, R. (2020). Ecological and economic benefits of low-intensity urban lawn management. *Journal of Applied Ecology, 57*(2), 436–446.

6

GARDENS AS THERAPY

We use the word *therapeutic* in two senses: first, as a colloquial way to express a general lift in mood; second, to refer to a formal treatment for a disease or dysfunction. In the first sense we may say that going into the garden to cut the grass is therapeutic as it provides exercise and the garden looks better afterwards. However, the informal use of the term contrasts with its use to refer to a recognised treatment for an established physical or psychological condition. While the focus here is specifically on gardens, there is a much wider literature on nature-based therapies – such as care farming (de Bruin et al., 2021), forest therapy (Zhang & Ye, 2022), therapeutic landscapes (Williams, 2010) and the effects of green spaces on health (Browning et al., 2022) – which are not included here (see Bonham-Corcoran et al., 2022). There is also some support for the wider claim of therapeutic benefits of contact with nature, which includes gardens and gardening (e.g., Seresinhe, Preis & Moat, 2015) alongside community gardens (e.g., Gregis et al., 2021), nature-based outdoor activities (Coventry et al., 2021), allotments (e.g., Genter et al., 2015), gardens built for therapeutic purposes such as healing gardens and gardens for horticultural therapy (Briggs, Morris, & Rees, in press; Spano et al., 2020) and indoor plants (Han & Ruan, 2019). Given the area of concern, gardens within institutions such as hospitals are included here alongside private gardens.

Health and Gardens

With a focus on gardens, Soga, Gaston and Yamaura (2017) carried out a meta-analysis of 22 studies from several countries of the effects of gardening on health. The study included both clinical and non-clinical populations and

DOI: 10.4324/9781003289661-7

was based on various indices of physical health such as bone mineral density, body mass index, fatigue and heart rate; alongside indicators of psychological health including anger, anxiety, depression, loneliness, self-esteem and stress. Soga et al. concluded that: "It is obvious that gardening has both immediate and long-term effects on health, and an important direction for future research is to determine the shape of relationships between the dose (duration and frequency) of gardening exercise health outcomes" (p. 97). This latter point was also made by Hartig et al. (2014) with the added caution that the gaps in knowledge indicate prudence in applying what is known.

de Bell et al. (2020) conducted a national survey in England ($n = 7,814$) of the effects on health – taken as evaluative wellbeing, general health, eudaimonic wellbeing and physical activity – of spending time in the garden. They reported that individuals with access to a private garden compared to those with no access to a private garden were significantly more likely to report high levels of evaluative and eudaimonic wellbeing, alongside good levels of physical activity. However, people with access to a communal garden had significantly *worse* reported general health than those with no garden. de Bell et al. suggest that privacy is the important factor: while there are socio-demographic variations in garden access given the cost of property with a private garden, privacy avoids the consensual necessities that are part of shared spaces. Thus, privacy allows the individual to create their garden in keeping with their own tastes and values, thereby making a clear distinction with public green spaces (Coolen & Meesters, 2012).

An Austrian study by Cervinka et al. (2016) reinforced the importance of privacy. They used the Perceived Restorativeness Scale (PRS; Hartig et al., 1997) to look at the effect of private gardens on wellbeing. The PRS assesses the individual's experience of the environment in terms of how much it provides a break from everyday life and the enjoyment it gives. Cervinka et al. found that: "All evaluated private spaces with green elements such as living rooms, balconies, terraces or gardens scored high on PRS, with the garden showing the highest restorative potential" (p. 185). Thus, a combination of the physical and social aspects of gardening can lift mood and help people to feel better about themselves (e.g., Chalmin-Pui et al., 2021; Zhang, Wang & Fang, 2022).

This chapter considers the evidence for the positive effects of gardens on physical and psychological health.

The interplay between health, gardens and gardening can be illustrated by reference to specific health concerns. As noted previously, there are a wide range of nature-based therapies. In considering the literature, it is not always possible to disentangle the specific effects of gardens and gardening. Another complicating issue in considering health, particularly mental health, lies in terminology. For example, a term such as *anxiety* may be used with clinical precision in keeping with a formal diagnostic system or alternatively as a synonym for feelings of unease or foreboding. This variation in usage makes it difficult to compare across studies and therefore reach solid conclusions.

Cancer

Phelps et al. (2015) report a study conducted in South Wales focused on seven women who had previously or recently had a diagnosis of breast cancer. The women planted an indoor garden bowl in their own home and cared for it over a three-month period. The women said that they found the experience therapeutic in that it prompted them to think about their health, triggered feelings of hope and in one case the motivation to engage in more gardening. Blaschke, O'Callaghan and Schofield (2017) contacted a range of experts and asked for their views on nature-based care of oncology patients. The experts broadly endorsed the beneficial effects of greenery through means such as positioning the patient's bed to view the outside world, taking patients outside when possible, and providing nature-based stimuli such as soundtracks and pictures of nature in the patient's room.

Cutillo et al. (2015) undertook a review of the research on nature-based therapy in cancer care, including 18 studies and making the distinction between *healing* gardens and *therapeutic* gardens: "While healing gardens are created for refuge and open to all visitors, therapeutic gardens are a form of therapy specifically aimed at healing sick individuals" (p. 4). They conclude that the evidence supports gardening as a complimentary therapy in medical settings for adults, with promise for paediatric populations awaiting further evidence.

Children's Health

As seen with gardens in schools, the received wisdom is that exposure to nature, including but not limited to gardens, is good for child development. Two systematic reviews have looked at the effects of nature, including school gardens, on children's health (Fyfe-Johnson et al., 2021; Sprague et al., 2022). They concluded that nature had significant benefits on physical activity as well as the child's mental health, behavioural and cognitive functioning. Tillmann et al. (2018) carried out a systematic review of the benefits of exposure to nature for the mental health for children and teenagers. They reached the conclusion "That interacting with nature is positively associated with the mental health of children and teenagers. The findings, although somewhat inconsistent and often non-significant, demonstrate the need for more in depth and rigorous research" (pp. 964–965).

In developmental psychology, a strong attachment between mother and child has long been held to be enormously beneficial for the child's development (Bowlby, 1953). Kotozaki (2020) engaged 15 postpartum women with infants under the age of 1 year to take part in a study that aimed to increase attachment. The mothers taking part in research, which used a pre-post design, worked on various gardening tasks such as planting and tending plants, weeding and gathering cut flowers. The infant was looked after by nursery staff while the mother gardened. Before and after the gardening the women provided information about their child's emotional state, their own attachment to their child, postpartum depression and stress due to parenting. Following the gardening the child's rhythmicity, a feature of infant emotion, showed a positive increase; there were improvements in mother–infant attachment, postpartum depression and parenting stress. The gardening benefited the mother and in turn their infant child.

A web-based American survey by Waliczek et al. (2000) asked 323 adults, parents and teachers, about what they saw as the benefits of gardening for the children they were actively involved with in school, community or domestic gardens. The adults reported that they were engaged in gardening with a total of 128,836 children, all under 18 years of age and with under 12s the largest group. The majority of children gardened in school, mainly growing fruits and vegetables. The adults felt that the gardening helped the child's self-esteem and reduced their stress levels. The parents and teachers did not agree on the most important aspect of the child's gardening; while parents thought growing food to be most important, teachers favoured socialising and increasing knowledge about plants.

A study in New Zealand by van Lier et al. (2017) drew on data from a 2012 national youth health and wellbeing survey provided by 8,500 randomly selected adolescent male and female students between 13 and 17 years of age. In all, one-quarter of the adolescents gardened at home. These gardeners were mainly younger males living in a rural area, characterised by healthy diets, higher levels of physical activity and low depressive symptomatology. When compared to non-gardeners, the gardeners showed greater consumption of fruits and vegetables and more physical activity, alongside improved wellbeing and mental health, particularly low levels of depression and heightened emotional wellbeing. The gardeners also reported a stronger connection to their family. Knoff et al. (2002) also found health benefits for adolescents with access to a community or home garden. Those adolescents with garden access ate more vegetables and rated themselves as healthier than those without access.

COVID-19

For a long period of time the COVID-19 pandemic changed the lives of millions of people. There were enforced periods of lockdown during which contact between family and friends was severely restricted as prescribed social distancing and self-isolation were used to attempt to curtail transmission. These restrictions on social activity, more pronounced for some sections of the population, had the potential to diminish health. A Scottish study by Corley et al. (2021) considered whether access to an allotment or a private garden was related to physical and mental wellbeing in older adults during the pandemic. In Scotland during the pandemic, it was strongly recommended that people aged over 70 years should stay at home during lockdown and avoid social contact. Corley et al. used extant data from the Lothian Birth Cohort 1936 study of ageing together witha self-report online survey to gather information on 171 people aged 84 years. The respondents were asked about their access to and use of a garden, their physical and mental health, and in retrospect whether their health had changed since lockdown. The respondents reported that compared with pre-lockdown they spent more time in the garden during lockdown while "Those participants reporting higher garden usage, compared with pre-lockdown, experienced more positive health outcomes" (p. 4).

Turnšek et al. (2022) conducted a survey across five European countries – Slovenia, Norway, Estonia, Switzerland and Iceland – to investigate if home gardening increased during the COVID pandemic. The responses from the 1,144 participants indicated the pandemic gave an impetus towards gardening at home. The move to the garden was in some instances a consequence of concerns about food supplies. The positive effects of gardening on health during the pandemic is evident on a global scale as seen, for example, from studies in Australia (Marsh et al., 2021), Brazil (Marques et al., 2021), India (Basu et al., 2021), Indonesia (Harding et al., 2022), Spain (Maury-Mora, Gómez-Villarino & Varela-Martínez, 2022), Sri Lanka (Weerakoon, Wehigaldeniya & Charikawickramarathne, 2021) and the United States (Gerdes et al., 2022).

Dementia

Dementia is a condition characterised by confusion, memory loss, poor judgement and wandering and becoming lost in familiar locations. The number of people in the United Kingdom with dementia is approaching one million. With age a significant risk factor, comparatively few young people develop the condition, but among people aged 65–69 years, about two in 100 people will have dementia. The risk then approximately doubles every 5 years so that amongst those over 90 years about 33 people in 100 will have dementia. Influenced by the severity of the condition, people with dementia may reside in their own home or live in a care home. A therapeutic garden (sometimes called a healing garden) is designed to help the individual to feel mentally and physically refreshed. A therapeutic garden may be specifically designed for people with certain conditions such as Alzheimer's disease, cancer or dementia (Thaneshwari, Sharma & Sahare, 2018). Motealleh et al. (2022) list eight attributes – accessibility, meaningful activity, orientation, reminiscence, safety, sensory stimulation, socialisation and sustainability – which should be incorporated into a dementia-friendly garden.

The effects of a therapeutic garden for people with dementia living in the community and in residential care has been extensively researched, and there are several reviews and systematic reviews (Barrett, Evans & Mapes, 2019; Murroni et al., 2021; Newton et al., 2021; Scott et al., 2022; Whear et al., 2014; Zhao, Liu & Wang, 2022). The conclusion from these reviews is mixed; for example, Scott et al. failed to find convincing evidence for an improvement in cognitive function, whereas Zhao et al. reported that horticultural therapy had a beneficial effect on cognitive function. The reasons for such inconsistencies, not uncommon in applied psychological research, are explicable by studies using different interventions with varying measures of outcome variables (e.g., cognition), differences in setting and participant sampling including type and severity of symptoms, and the use of weaker research designs. These issues are recognised in the reviews alongside appeals for large-scale, high-quality studies to allow more robust conclusions to be formulated.

Elderly Ailments

Growing older brings about significant changes in physical and psychological health alongside the additional risk of conditions such as dementia. There are arguments in favour of nature-based activities, such as horticultural therapy, gardening and visiting green spaces for improving the wellbeing of elderly people (Gagliardi & Piccinini, 2019; Lin et al., 2022; Sia et al., 2020). However, as Nicholas, Giang and Yap (2019) point out, while horticultural therapy is widely used with older adults, it is difficult to be certain of its effects. Although the evidence looks hopeful, the variations in practice, both in therapeutic content and quantity, make it impossible to draw firm conclusions about effectiveness (Detweiler et al., 2012).

When gardening is the specific focus, there is a substantial international literature pointing to the physical, psychological and social benefits of gardening for older people (e.g., Kim et al., 2020; Scott, Masser & Pachana, 2015). Wang and MacMillan (2013) carried out a systematic review of the literature, including 22 studies, of the benefits of gardening for older people. They reached the view that: "Overall, the majority of the studies found some evidence that gardening is enjoyable for older adults and that it benefits overall quality of life, physical ability, and activeness" (p. 175). The gardeners identified intellectual stimulation and creativity among the benefits of gardening, alongside remembering previous family experiences and connecting to nature. There were variations within this overall picture; for those people living in the community, gardening had a greater influence on health and wellbeing than for those living in institutions.

Many gardeners continue to garden as they grow older, adjusting their routines to accommodate their frailties. While the exercise provided by gardening may have beneficial effects on wellbeing, including lowered mortality (Lêng & Wang, 2016), ideally gardening should be part of a network of complementary fitness and social activities (Glass et al., 1999). An Australian study by Scott, Masser and Pachana (2015) surveyed 331 adults, predominately female and retired aged from 60 to 90 years old, focusing on what they saw as the benefits of gardening. As they grew older, the participants adapted their gardening techniques; they gardened more frequently but for a shorter period of time; planned in advance how to complete a gardening job; moved to a house with a smaller garden; used tools adapted to their needs; and sought help, either from friends and family or from paid workers, for heavy tasks such as digging. The reasons they gave for continuing to garden highlighted the value they placed on the beauty of gardens and their link with nature, as well as a sense of achievement following their mental and physical activities. Scott, Masser and Pachana suggest that for their respondents gardening is more than a casual leisure pursuit; it makes a significant contribution to their physical and psychological wellbeing.

The physical limitations accompanying aging may result in a gardener becoming unable to engage in an activity that not only gives them pleasure but is an integral part of their identity (Cheng & Pegg, 2016). Some older people compensate for the loss of a valued activity, as well as alleviate worries about the lack of maintenance, by employing a gardening service. An Australian study by Same et al. (2016) gathered information from elderly people in receipt of an inexpensive gardening service. As well as keeping the garden tidy, the service looks to safety and therefore clears pathways, prunes overhanging branches, mows lawns and clears green waste. The elderly person is encouraged to carry out less strenuous gardening jobs such as watering and light pruning. Same et al. reported that the elderly people said they worried less about their garden's upkeep and appreciated the safety aspects such as the reduced chance of a fall. Thus, the gardening service has the additional benefit of helping the elderly people to feel safe to stay in their own homes, which is in keeping with the wishes of the majority of this segment of the population. In all, as Same et al. state: "The well-maintained garden also promoted dignity and self-esteem for the participants as they commented that the garden was a reflection and expression of themselves. Gardens are indeed connected with the home owners' identities, reflecting their culture, gender, and social class" (p. 260).

Zhang, Wang and Fang (2022) reported a systematic review and meta-analysis of 13 studies concerned with the effects of various types of horticultural therapy on depressive symptoms in elderly people. The studies included gardening, which was widely practiced, as well as other interventions, including nature–art activities such as flower arranging and field trips to green environments. Although the studies assessed depressive symptoms using a range of methods, the overall conclusion was that the interventions had a significant positive effect on reducing depressive symptoms. However, a thorough examination of the evidence relating to gardening and physical health among the elderly led Nicklett, Anderson and Yen (2016) to conclude that "The current literature does not provide sufficient evidence of the physical functioning consequences of gardening" (p. 678). The reasons for these apparently contradictory conclusions are discussed in Chapter 7.

Mental Health and General Wellbeing

Mental health is a generic, catch-all term that is used both with reference to specific conditions such as depression and anxiety disorders as well to those indefinable feelings and behaviours that give us concern. *Wellbeing* is another catch-all term by which we refer to our subjective state of physical or psychological health. As detailed by Taylor et al. (2022), wellbeing can be assessed by interview, formal rating scales, questionnaire and self-report. There are studies given to horticultural therapy for mental health generally (Tu, 2022) and specific conditions such as depression (e.g., Chen, 2021) and schizophrenia (e.g., Lu et al., 2021), but this therapeutic approach is broader in scope than gardening.

Clatworthy, Hinds and Camic (2013) reviewed ten research studies that evaluated gardening-based mental health interventions for a range of symptoms, including anxiety and depression. The participants in the studies variously described the emotional, physical, social, spiritual and vocational benefits that followed treatment. Clatworthy, Hinds and Camic make two comments on the quality of the research published over the past two decades. First, satisfactory outcome measures are needed to allow greater accuracy in measurement and to increase replicability. Second, none of the studies reviewed employed a randomised control trial research design, which leaves open the need for higher quality research. A further issue, highlighted by Ainamani et al. (2022), is a scarcity of research looking at gardening and mental health carried out in low- and middle-income countries.

Gardening in an allotment may bring a social element not found in a private garden. Soga et al. (2017) carried out a questionnaire survey of 332 people in Tokyo comparing aspects of self-reported health – body mass index, general health, subjective health complaints, mental health and social cohesion – between allotment gardeners and non-gardeners. The gardeners self-reported better health on all the indices except of body mass index. An interesting finding was that for the gardeners neither frequency nor duration of gardening was significantly related to self-reported health. Thus, those who gardened less often and for shorter periods of time self-reported similar health benefits. While exercising due caution in over-elaborating from a single study, although a British study recorded a not dissimilar finding (Wood, Pretty & Griffin, 2016), it may be that the psychological consequences of gardening outweigh the physical in influencing self-perception of health. In other words, we expect to feel better for gardening and so we report that, indeed, we are. Although, to complicate matters further, gardeners' self-reported heath may vary as a function of garden size (Brindley, Jorgensen & Maheswaran, 2018). A study employing objective measures of health would help advance this line of enquiry.

Briggs, Morris and Rees (in press) consider the social dimension to gardening in their review of 24 studies of group-based gardening intended to improve wellbeing and reduce symptoms of mental illness. They suggest that the indications are positive with regard to improving wellbeing and reducing indicators of depression. However, there are issues with the evidence base caused by small samples, the risk of bias, and inadequate reporting.

Stress

The difficulty noted above regarding terminology is immediately apparent when it comes to stress. In their review of nature-based interventions for stress-related illness Johansson et al. (2022) employed a range of terms drawn from the literature, including *burnout, exhaustion disorder, psychological stress, occupational stress* and *stress*. Although some diagnostic systems have formal definitions of stress-related disorders (e.g., DSM-5; American Psychiatric Association, 2013), the varied use of terms should nonetheless be kept in mind when considering the effects of gardening (Figure 6.1).

Adevi and Mårtensson (2013) investigated the potential of the garden to aid recovery from stress. Five individuals on leave from work because of a stress-related problem were interviewed about their experience at the Alnarp Rehabilitation Garden in southern Sweden. The individuals were positive about interacting with other patients and carers in what they saw as a green and safe environment. They said that while mainstream therapy had initiated their recovery, access to the garden had strengthened the process. Johansson, Juuso and Engström (2022) reviewed 25 studies concerned with forest (5 studies) or garden (20 studies) interventions for people with a stress-related illness. They concluded that the interventions effectively reduced stress and promoted physical and psychological health, enabling participants to return to work.

FIGURE 6.1 Garden Meditation.

Source: Photographer Jared Rice.

On a much more expansive basis, the garden can provide stress relief during a pandemic. Kingsley et al. (2022) carried out a large international survey asking gardeners about their experiences in 2020 when the COVID-19 pandemic was emerging. They found that as the pandemic unfolded, gardening formed a safe space for social contact with family and friends. In a multinational survey, Egerer et al. (2022) found that gardeners said their garden had many social and psychological benefits, including stress reduction and providing fruits and vegetables, during the pandemic.

Judging by the extant evidence, there are grounds for supposing that the act of gardening can in itself bring about a reduction in stress. An innovative study by Stigsdotter et al. (2018) compared the effects of cognitive behaviour therapy (CBT), a validated method of treatment, with a nature-based therapy (NBT) based on a combination of psychological techniques such as mindfulness alongside a range of individual gardening activities. They used a Randomised Control Trial, taken to be the optimal design in this type of research, with a burnout questionnaire and scores on measures of psychological wellbeing as the outcome variables. They found that both treatments had significant positive outcomes that remained at a 1-year follow-up; there was no significant difference in effectiveness between the two types of treatment. However, as Coventry and White (2018) point out, a single study cannot make the case that nature plays a role in managing stress-related problems. The relative contributions of exposure to nature and CBT are not known and, in addition, the characteristics of the participants in the study may have influenced the outcome.

Hospital Gardens

There are several types of hospitals for people with a variety of physical and mental health conditions. Some people go into hospital for a short time, others for much longer, and some never leave. The architectural design of hospitals must meet several complex and inter-related demands: (i) the physical arrangement of wards, corridor design and location of staff stations can have an effect on professional care and patient behaviour (Bernhardt et al., 2022; Moslehian et al., 2022); and (ii) the design of the patient's immediate environment may facilitate their comfort and healing (Simonsen, Sturge & Duff, 2022). Gardens play an important role in hospital design. Thus, while windows can make a hospital room feel bright and airy, a view of nature through a hospital window can facilitate a patient's recovery (Ulrich, 1984). Gardens within a hospital can be of benefit for both patients and their visitors (Martin et al., 2021; Untaru et al., 2022; Whitehouse et al., 2001).

If the interplay between gardens and health really is beneficial to those with ill health, then is it realistic to suggest that patients could be prescribed time in the garden to improve their health (Coventry & White, 2018)?

Prescribing Green Cures

The move in many countries to add *social prescribing* to traditional medical procedures aims to enable individuals and communities to improve their health and wellbeing (Cuthbert, Kellas & Page, 2021; Morse et al., 2022). Social prescribing may incorporate green, natural elements including gardens into patient care (Howarth et al., 2020) and is prevalent in the United Kingdom (Moore et al., 2022; Robinson et al., 2020). The aim of green prescribing is to connect the patient with nature using methods such as conservation, care farming, gardening, horticulture, walking groups and woodland groups. These activities may include social interaction between patients and between patients and carers.

The broad question is "does social prescribing work?"; the narrower question of greater interest here is "does *green* social prescribing work?" In trying to answer these questions, several fundamental issues are prevalent. As Elliott et al. (2022) state:

> There is no agreed definition of social prescribing, but it is generally understood to involve referral to non-medical resources in the community, with the goal of improved health and well-being. This typically involves a link worker, also known as a community connector or navigator, who works with the individual to identify their needs, coproduce goals and connect them to resources in their community. (p. 1)

The definitional ambiguities highlighted by Elliott et al. make evaluation a difficult task. Yet further, the practical issues involved are highlighted in a study by Wood et al. (2022) looking at a therapeutic community gardening project prescribed for mental illness. Wood et al. interviewed 13 people involved in administration of the project, such as link workers and garden staff, and held focus groups with 20 garden members. The aspects of the scheme, which aided the project's referral, uptake and attendance, were seen as a person-centred approach flexible to individual health needs. The barriers to referral included a lack of awareness of the full range of therapeutic community gardens and their physical location. Wood et al. argue that nature-based interventions engage participants and improve health through a connection with nature and the outdoors, through providing hope and by supporting social relationships. In addition, they suggest that this green approach makes financial savings by reducing reliance on other health services.

How robust is the evidence in support of the claims for the benefits, both to health and public finances, of green prescribing? While not dismissing the idea of social prescribing, more persuasive evidence in support of its efficacy is clearly required (Husk et al., 2019). There are several reviews of green prescribing that have the consistent message of potential gains from social prescribing but are allied to the need for strong supporting evidence (Bonham-Corcoran, 2022; Elliott et al., 2022; Masterton et al., 2020; Nguyen et al., 2022). In addition, Buckley and Chauvenet (2022) show that contact with nature can be highly cost effective in reducing healthcare costs and maintaining worker productivity.

There may be a case to be made for the benefits to health of time spent in the garden. However, if this possibility is to assume a more formal footing in healthcare services, then the evidence on which the assumption rests should meet certain standards. The next chapter discusses the importance of these standards, the foundations on which they are built, and the realities of providing evidence in their support.

References

Adevi, A. A., & Mårtensson, F. (2013). Stress rehabilitation through garden therapy: The garden as a place in the recovery from stress. *Urban Forestry & Urban Greening*, *12*(2), 230–237.

American Psychiatric Association. (2013). *Diagnostic and statistical manual of mental disorders*, 5. Arlington, VA: Author.

Ainamani, H. E., Gumisiriza, N., Bamwerinde, W. M., & Rukundo, G. Z. (2022). Gardening activity and its relationship to mental health: Understudied and untapped in low-and middle-income countries. *Preventive Medicine Reports*, *29*, 101946.

Barrett, J., Evans, S., & Mapes, N. (2019). Green dementia care in accommodation and care settings: A literature review. *Housing, Care and Support*, *22*(4), 193–206.

Basu, M., DasGupta, R., Kumar, P., & Dhyani, S. (2021). Home gardens moderate the relationship between covid-19-induced stay-at-home orders and mental distress: A case study with urban residents of India. *Environmental Research Communications*, *3*(10), 105002.

Bernhardt, J., Lipson-Smith, R., Davis, A., White, M., Zeeman, H., Pitt, N., ... & Elf, M. (2022). Why hospital design matters: A narrative review of built environments research relevant to stroke care. *International Journal of Stroke*, *17*(4), 370–377.

Blaschke, S., O'Callaghan, C. C., & Schofield, P. (2017). Nature-based care opportunities and barriers in oncology contexts: A modified international e-Delphi survey. *BMJ Open*, *7*(10), e017456.

Bonham-Corcoran, M., Solovyeva, A., O'Briain, A., Cassidy, A., & Turner, N. (2022). The benefits of nature-based therapy for the individual and the environment: An integrative review. *Irish Journal of Occupational Therapy*, *50*(1), 16–27.

Bowlby, J. (1953). *Child care and the growth of love*. Harmondsworth, Middlesex: Penguin.

Briggs, R., Morris, P. G., & Rees, K. (in press). The effectiveness of group-based gardening interventions for improving wellbeing and reducing symptoms of mental ill-health in adults: A systematic review and meta-analysis. *Journal of Mental Health*.

Brindley, P., Jorgensen, A., & Maheswaran, R. (2018). Domestic gardens and self-reported health: A national population study. *International Journal of Health Geographics*, *17*(1), 1–11.

Browning, M. H. E. M., Rigolon, A., McAnirlin, O., & Yoon, H. (2022). Where greenspace matters most: A systematic review of urbanicity, greenspace, and physical health. *Landscape and Urban Planning*, *217*, 104233.

Buckley, R. C., & Chauvenet, A. L. (2022). Economic value of nature via healthcare savings and productivity increases. *Biological Conservation*, *272*, 109665.

Cervinka, R., Schwab, M., Schönbauer, R., Hämmerle, I., Pirgie, L., & Sudkamp, J. (2016). My garden–my mate? Perceived restorativeness of private gardens and its predictors. *Urban Forestry & Urban Greening*, *16*, 182–187.

Chalmin-Pui, L. S., Roe, J., Griffiths, A., Smyth, N., Heaton, T., Clayden, A., & Cameron, R. (2021). "It made me feel brighter in myself" – The health and well-being impacts of a residential front garden horticultural intervention. *Landscape and Urban Planning*, 205, 103958.

Chen, H. (2021). The effect of horticultural therapy in depression intervention. *Journal of Landscape Research*, 13(6), 13–22.

Cheng, E., & Pegg, S. (2016). "If I'm not gardening, I'm not at my happiest": Exploring the positive subjective experiences derived from serious leisure gardening by older adults. *World Leisure Journal*, 58(4), 285–297.

Clatworthy, J., Hinds, J., Camic, P. M. (2013). Gardening as a mental health intervention: A review. *Mental Health Review Journal*, 18(4), 214–225.

Coolen, H., & Meesters, J. (2012). Private and public green spaces: Meaningful but different settings. *Journal of Housing and the Built Environment*, 27, 49–67.

Corley, J., Okely, J. A., Taylor, A. M., Page, D., Welstead, M., Skarabela, B., ... & Russ, T. C. (2021). Home garden use during COVID-19: Associations with physical and mental wellbeing in older adults. *Journal of Environmental Psychology*, 73, 101545.

Coventry, P. A., Brown, J. E., Pervin, J., Brabyn, S., Pateman, R., Breedvelt, J., ... & White, P. L. (2021). Nature-based outdoor activities for mental and physical health: Systematic review and meta-analysis. *SSM-Population Health*, 16, 100934.

Coventry, P. A., & White, P. C. (2018). Are we ready to use nature gardens to treat stress-related illnesses? *The British Journal of Psychiatry*, 213(1), 396–397.

Cuthbert, S., Kellas, A., & Page, L. A. (2021). Green care in psychiatry. *The British Journal of Psychiatry*, 218(2), 73–74.

Cutillo, A., Rathore, N., Reynolds, N., Hilliard, L., Haines, H., Whelan, K., & Madan-Swain, A. (2015). A literature review of nature-based therapy and its application in cancer care. *Journal of Therapeutic Horticulture*, 25(1), 3–15.

de Bell, S., White, M., Griffiths, A., Darlow, A., Taylor, T., Wheeler, B., & Lovell, R. (2020). Spending time in the garden is positively associated with health and wellbeing: Results from a national survey in England. *Landscape and Urban Planning*, 200, 103836.

de Bruin, S., Hassink, J., Vaandrager, L., de Boer, B., Verbeek, H., Pedersen, I., ... & Eriksen, S. (2021). Care farms: A health-promoting context for a wide range of client groups. In E. Brymer, M. Rogerson, & J. Barton (Eds.), *Nature and Health: Physical Activity in Nature* (pp. 177–190). New York: Routledge.

Detweiler, M. B., Sharma, T., Detweiler, J. G., Murphy, P. F., Lane, S., Carman, J., ... & Kim, K. Y. (2012). What is the evidence to support the use of therapeutic gardens for the elderly? *Psychiatry Investigation*, 9(2), 100–110.

Elliott, M., Davies, M., Davies, J., & Wallace, C. (2022). Exploring how and why social prescribing evaluations work: A realist review. *BMJ Open*, 12(4), e057009.

Egerer, M., Lin, B., Kingsley, J., Marsh, P., Diekmann, L., & Ossola, A. (2022). Gardening can relieve human stress and boost nature connection during the COVID-19 pandemic. *Urban Forestry & Urban Greening*, 68, 127483.

Fyfe-Johnson, A. L., Hazlehurst, M. F., Perrins, S. P., Bratman, G. N., Thomas, R., Garrett, K. A., ... & Tandon, P. S. (2021). Nature and children's health: A systematic review. *Pediatrics*, 148(4), e2020049155.

Gagliardi, C., & Piccinini, F. (2019). The use of nature-based activities for the well-being of older people: An integrative literature review. *Archives of Gerontology and Geriatrics*, 83, 315–327.

Genter, C., Roberts, A., Richardson, J., & Sheaff, M. (2015). The contribution of allotment gardening to health and wellbeing: A systematic review of the literature. *British Journal of Occupational Therapy, 78*(10), 593–605.

Gerdes, M. E., Aistis, L. A., Sachs, N. A., Williams, M., Roberts, J. D., & Rosenberg Goldstein, R. E. (2022). Reducing anxiety with nature and gardening (RANG): Evaluating the impacts of gardening and outdoor activities on anxiety among US adults during the Covid-19 pandemic. *International Journal of Environmental Research and Public Health, 19*(9), 5121.

Glass, T. A., De Leon, C. M., Marottoli, R. A., & Berkman, L. F. (1999). Population based study of social and productive activities as predictors of survival among elderly Americans. *British Medical Journal, 319*(7208), 478–483.

Gregis, A., Ghisalberti, C., Sciascia, S., Sottile, F., & Peano, C. (2021). Community garden initiatives addressing health and well-being outcomes: A systematic review of infodemiology aspects, outcomes, and target populations. *International Journal of Environmental Research and Public Health, 18*(4), 1943.

Han, K. T., & Ruan, L. W. (2019). Effects of indoor plants on self-reported perceptions: A systemic review. *Sustainability, 11*(16), 4506.

Harding, D., Lukman, K. M., Jingga, M., Uchiyama, Y., Quevedo, J. M. D., & Kohsaka, R. (2022). Urban gardening and wellbeing in pandemic era: Preliminary results from a socio-environmental factors approach. *Land, 11*(4), 492.

Hartig, T., Mitchell, R., De Vries, S., & Frumkin, H. (2014). Nature and health. *Annual Review of Public Health, 35*, 207–228.

Hartig, T., Korpela, K., Evans, G. W., & Gärling, T. (1997). A measure of restorative quality in environments. *Scandinavian Housing and Planning Research, 14*(4), 175–194.

Howarth, M., Griffiths, A., da Silva, A., & Green, R. (2020). Social prescribing: A 'natural' community-based solution. *British Journal of Community Nursing, 25*(6), 294–298.

Husk, K., Elston, J., Gradinger, F., Callaghan, L., & Asthana, S. (2019). Social prescribing: Where is the evidence? *British Journal of General Practice, 69*(678), 6–7.

Johansson, G., Juuso, P., & Engström, Å. (2022). Nature-based interventions to promote health for people with stress-related illness: An integrative review. *Scandinavian Journal of Caring Sciences, 33*(4), 910–925.

Kim, Y. H., Lee, S. H., Park, C. S., Bae, H. O., Kim, Y. J., & Huh, M. R. (2020). A horticultural therapy program focusing on gardening activities to promote psychological, emotional and social health of the elderly living in a homeless living facility for a long time: A pilot study. *Journal of People, Plants, and Environment, 23*(5), 565–576.

Kingsley, J., Diekmann, L., Egerer, M. H., Lin, B. B., Ossola, A., & Marsh, P. (2022). Experiences of gardening during the early stages of the COVID-19 pandemic. *Health & Place, 76*, 102854.

Knoff, K. A., Kulik, N., Mallare, J., & Dombrowski, R. D. (2022). The association between home or community garden access and adolescent health. *Family & Community Health, 45*(4), 267–271.

Kotozaki, Y. (2020). Psychological effects of the gardening activity on mother and their infant: Preliminary evidence from an exploratory pilot study. *Psychology, 11*(9), 1349–1360.

Lêng, C. H., & Wang, J. D. (2016). Daily home gardening improved survival for older people with mobility limitations: An 11-year follow-up study in Taiwan. *Clinical Interventions in Aging, 11*, 947.

Lin, Y., Lin, R., Liu, W., & Wu, W. (2022). Effectiveness of horticultural therapy on physical functioning and psychological health outcomes for older adults: A systematic review and meta-analysis. *Journal of Clinical Nursing, 31*(15–16), 2087–2099.

Lu, S., Zhao, Y., Liu, J., Xu, F., & Wang, Z. (2021). Effectiveness of horticultural therapy in people with schizophrenia: A systematic review and meta-analysis. *International Journal of Environmental Research and Public Health, 18*(3), 964.

Marques, P., Silva, A. S., Quaresma, Y., Manna, L. R., de Magalhães Neto, N., & Mazzoni, R. (2021). Home gardens can be more important than other urban green infrastructure for mental well-being during COVID-19 pandemics. *Urban Forestry & Urban Greening, 64*, 127268.

Marsh, P., Diekmann, L. O., Egerer, M., Lin, B., Ossola, A., & Kingsley, J. (2021). Where birds felt louder: The garden as a refuge during COVID-19. *Wellbeing, Space and Society, 2*, 100055.

Martin, K., Nanu, L., Kwon, W. S., & Martin, D. (2021). Small garden, big impact: Emotional and behavioral responses of visitors to a rooftop atrium in a major hospital. *HERD: Health Environments Research & Design Journal, 14*(3), 274–287.

Masterton, W., Carver, H., Parkes, T., & Park, K. (2020). Greenspace interventions for mental health in clinical and non-clinical populations: What works, for whom, and in what circumstances? *Health & Place, 64*, 102338.

Maury-Mora, M., Gómez-Villarino, M. T., & Varela-Martínez, C. (2022). Urban green spaces and stress during COVID-19 lockdown: A case study for the city of Madrid. *Urban Forestry & Urban Greening, 69*, 127492.

Moore, C., Unwin, P., Evans, N., & Howie, F. (2022). Social prescribing: Exploring general practitioners' and healthcare professionals' perceptions of, and engagement with, the NHS model. *Health & Social Care in the Community,30*(6), e5176-e5185.

Morse, D. F., Sandhu, S., Mulligan, K., Tierney, S., Polley, M., Giurca, B. C., … & Husk, K. (2022). Global developments in social prescribing. *BMJ Global Health, 7*(5), e008524.

Moslehian, A. S., Tucker, R., Kocaturk, T., & Andrews, F. (2022). Analysis of factors historically affecting innovation in hospital building design. *HERD: Health Environments Research & Design Journal, 15*(4), 249–269.

Motealleh, P., Moyle, W., Jones, C., & Dupre, K. (2022). The impact of a dementia-friendly garden design on people With dementia in a residential aged care facility: A case study. *HERD: Health Environments Research & Design Journal, 15*(2), 196–218.

Murroni, V., Cavalli, R., Basso, A., Borella, E., Meneghetti, C., Melendugno, A., & Pazzaglia, F. (2021). Effectiveness of therapeutic gardens for people with dementia: A systematic review. *International Journal of Environmental Research and Public Health, 18*(18), 9595.

Newton, R., Keady, J., Tsekleves, E., & Obe, S. A. (2021). 'My father is a gardener … ': A systematic narrative review on access and use of the garden by people living with dementia. *Health & Place, 68*, 102516.

Nguyen, P. Y., Rahimi-Ardabili, H., Feng, X., & Astell-Burt, T. (2022). Nature prescriptions: A scoping review with a nested meta-analysis. *medRxiv.*

Nicholas, S. O., Giang, A. T., & Yap, P. L. (2019). The effectiveness of horticultural therapy on older adults: A systematic review. *Journal of the American Medical Directors Association, 20*(10), 1351.e1–1351.e11.

Nicklett, E. J., Anderson, L. A., & Yen, I. H. (2016). Gardening activities and physical health among older adults: A review of the evidence. *Journal of Applied Gerontology, 35*(6), 678–690.

Phelps, C., Butler, C., Cousins, A., & Hughes, C. (2015). Sowing the seeds or failing to blossom? A feasibility study of a simple ecotherapy-based intervention in women affected by breast cancer. *Ecancermedicalscience, 9*, 602.

Robinson, J. M., Jorgensen, A., Cameron, R., & Brindley, P. (2020). Let nature be thy medicine: A socioecological exploration of green prescribing in the UK. *International Journal of Environmental Research and Public Health, 17*(10), 3460.

Same, A., Lee, E. A. L., McNamara, B., & Rosenwax, L. (2016). The value of a gardening service for the frail elderly and people with a disability living in the community. *Home Health Care Management & Practice, 28*(4), 256–261.

Scott, T. L., Jao, Y. L., Tulloch, K., Yates, E., Kenward, O., & Pachana, N. A. (2022). Well-being benefits of horticulture-based activities for community dwelling people with dementia: A systematic review. *International Journal of Environmental Research and Public Health, 19*(17), 10523.

Scott, T. L., Masser, B. M., & Pachana, N. A. (2015). Exploring the health and wellbeing benefits of gardening for older adults. *Ageing & Society, 35*(10), 2176–2200.

Seresinhe, C. I., Preis, T., & Moat, H. S. (2015). Quantifying the impact of scenic environments on health. *Scientific Reports, 5*(1), 1–9.

Sia, A., Tam, W. W., Fogel, A., Kua, E. H., Khoo, K., & Ho, R. (2020). Nature-based activities improve the well-being of older adults. *Scientific Reports, 10*(1), 1–8.

Simonsen, T., Sturge, J., & Duff, C. (2022). Healing architecture in healthcare: A scoping review. *HERD: Health Environments Research & Design Journal, 15*(3), 315–328.

Soga, M., Cox, D. T., Yamaura, Y., Gaston, K. J., Kurisu, K., & Hanaki, K. (2017). Health benefits of urban allotment gardening: Improved physical and psychological well-being and social integration. *International Journal of Environmental Research and Public Health, 14*(1), 71.

Spano, G., D'Este, M., Giannico, V., Carrus, G., Elia, M., Lafortezza, R., ... & Sanesi, G. (2020). Are community gardening and horticultural interventions beneficial for psychosocial well-being? A meta-analysis. *International Journal of Environmental Research and Public Health, 17*(10), 3584.

Soga, M., Gaston, K. J., & Yamaura, Y. (2017). Gardening is beneficial for health: A meta-analysis. *Preventive Medicine Reports, 5*, 92–99.

Sprague, N. L., Bancalari, P., Karim, W., & Siddiq, S. (2022). Growing up green: A systematic review of the influence of greenspace on youth development and health outcomes. *Journal of Exposure Science & Environmental Epidemiology, 32*, 1–22.

Stigsdotter, U. K., Corazon, S. S., Sidenius, U., Nyed, P. K., Larsen, H. B., & Fjorback, L. O. (2018). Efficacy of nature-based therapy for individuals with stress-related illnesses: Randomised controlled trial. *The British Journal of Psychiatry, 213*(1), 404–411.

Taylor, E. M., Robertson, N., Lightfoot, C. J., Smith, A. C., & Jones, C. R. (2022). Nature-based interventions for psychological wellbeing in long-term conditions: A systematic review. *International Journal of Environmental Research and Public Health, 19*(6), 3214.

Thaneshwari, P. K., Sharma, R., & Sahare, H. A. (2018). Therapeutic gardens in healthcare: A review. *Annals of Biology, 34*(2), 162–166.

Tillmann, S., Tobin, D., Avison, W., & Gilliland, J. (2018). Mental health benefits of interactions with nature in children and teenagers: A systematic review. *Journal of Epidemiology and Community Health, 72*(10), 958–966.

Tu, H. M. (in press). Effect of horticultural therapy on mental health: A meta-analysis of randomized controlled trials. *Journal of Psychiatric and Mental Health Nursing, 29*(4), 603–615.

Turnšek, M., Gangenes Skar, S. L., Piirman, M., Thorarinsdottir, R. I., Bavec, M., & Junge, R. (2022). Home gardening and food security concerns during the COVID-19 pandemic. *Horticulturae, 8*(9), 778.

Ulrich, R. S. (1984). View through a window may influence recovery from surgery. *Science, 224*(4647), 420–421.

Untaru, E. N., Ariza-Montes, A., Kim, H., & Han, H. (2022). Green environment, mental health, and loyalty among male and female patients. *Journal of Men's Health, 18*(10), 207.

van Lier, L. E., Utter, J., Denny, S., Lucassen, M., Dyson, B., & Clark, T. (2017). Home gardening and the health and well-being of adolescents. *Health Promotion Practice, 18*(1), 34–43.

Waliczek, T. M., Lineberger, R. D., Zajicek, J. M., & Bradley, J. C. (2000). Using a web-based survey to research the benefits of children gardening. *HortTechnology, 10*(1), 71–76.

Wang, D., & MacMillan, T. (2013). The benefits of gardening for older adults: A systematic review of the literature. *Activities, Adaptation & Aging, 37*(2), 153–181.

Weerakoon, R., Wehigaldeniya, D. S. & Charikawickramarathne. (2021). The effect of home gardening towards the mental well-being of men and women during the COVID-19 pandemic. *Asian Journal of Advances in Research, 10*(2), 1–8.

Whear, R., Coon, J. T., Bethel, A., Abbott, R., Stein, K., & Garside, R. (2014). What is the impact of using outdoor spaces such as gardens on the physical and mental well-being of those with dementia? A systematic review of quantitative and qualitative evidence. *Journal of the American Medical Directors Association, 15*(10), 697–705.

Whitehouse, S., Varni, J. W., Seid, M., Cooper-Marcus, C., Ensberg, M. J., Jacobs, J. R., & Mehlenbeck, R. S. (2001). Evaluating a children's hospital garden environment: Utilization and consumer satisfaction. *Journal of Environmental Psychology, 21*(3), 301–314.

Williams, A. (2010). Spiritual therapeutic landscapes and healing: A case study of St. Anne de Beaupre, Quebec, Canada. *Social Science & Medicine, 70*(10), 1633–1640.

Wood, C. J., Polley, M., Barton, J. L., & Wicks, C. L. (2022). Therapeutic community gardening as a green social prescription for mental ill-health: Impact, barriers, and facilitators from the perspective of multiple stakeholders. *International Journal of Environmental Research and Public Health, 19*(20), 13612.

Wood, C. J., Pretty, J., & Griffin, M. (2016). A case–control study of the health and well-being benefits of allotment gardening. *Journal of Public Health*, *38*(3), e336–e344.

Zhang, Y., Wang, J., & Fang, T. (2022). The effect of horticultural therapy on depressive symptoms among the elderly: A systematic review and meta-analysis. *Frontiers in Public Health*, *10*, 953363.

Zhao, Y., Liu, Y., & Wang, Z. (2022). Effectiveness of horticultural therapy in people with dementia: A quantitative systematic review. *Journal of Clinical Nursing*, *31*(13–14), 1983–1997.

Zhang, Z., & Ye, B. (2022). Forest therapy in Germany, Japan, and China: Proposal, development status, and future prospects. *Forests*, *13*(8), 1289.

7

ARE GARDENS REALLY THERAPEUTIC?

In the previous chapter the distinction was made between using the word *therapeutic* in a colloquial sense to express a general lift in mood as opposed to referring to a formal treatment for a diagnosable medical condition. The focus in this chapter is on the question of whether gardening actually has an identifiable therapeutic effect on our physical or mental health. When considered in relation to the colloquial use of the term, this question is really of no great importance. We feel a bit down in the dumps, so we go into the garden, take some exercise in the fresh air, and feel better for it. Although of benefit to the individual concerned, it has no wider, societal impact. The opposite is true when claims are made for the use of gardening as a therapy for medical conditions to be used with significant numbers of people. In the United Kingdom, if a medical intervention is to be approved for use with the public, then three safeguards must be satisfied. First, the individual must be protected from ineffective and dangerous treatments. Second, medical practitioners must be assured that they are using effective remedies for those in their care. Third, public money must be spent effectively with minimum wastage and maximum effectiveness. As discussed below, there are stringent standards of evidence that a medicine must satisfy before entering use. However, as Johnson (1999) states: "Plant–people interactions result in psychological benefits for the individual. The documentation supporting this assertion is extensive, though many studies may lack scientific validity" (p. 231).

The Importance of Scientific Evidence

There are many areas where people have different, often strongly held, opinions about what is best in life. These opinions range from the

DOI: 10.4324/9781003289661-8

mundane, say how best to make a cup of tea, to altogether more serious issues such as whether hitting children is an effective means of parenting. In cases where a policy is to be written or a new practise introduced to change people's lives, this is best done on the basis of evidence rather than rhetoric. There are many ways to gather evidence to inform practise. Qualitative research uses people's spoken words, say in an interview or a focus group, to gather information. Quantitative research applies a range of techniques, generally including a large number of respondents, such as opinion polls, questionnaires and experimental studies. Qualitative research has an emphasis on objective measurement alongside statistical analyses of fresh or archival data.

There are treatments available that sit outside the scientific and regulatory processes; these are generally referred to as *complementary* or *alternative* medicine. Some alternative medicines, such as homoeopathy, are widely accessible and used by many people who speak favourably of their effects. However, despite considerable research, the evidence base for the use of alternative medicines is lacking (e.g., Ernst, 2002). In this light, Masterton et al. (2020) make the case for the importance of evidence informing "green" clinical practise.

> With the increase in awareness of the benefits of being outside for mental health, more greenspace programmes are embedding mental health outcomes into their aims. This increases the risk that some programmes could be claiming all types of benefits, with little evidence to support claims. Without clarity of what approaches may or may not consist of, it is difficult to distinguish practice that is ethical and effective, from programmes that over-claim benefit and put users at risk of potential harm. This potentially makes it difficult to know which programmes to enrol on, or which programmes care providers should recommend. (p. 15)

As noted above, there are several reasons why policy and practise should be informed by a strong body of evidence, which include reducing the likelihood of harm and improving outcomes. In addition, the process of continually gathering evidence refines practise and promotes the quality, efficacy and cost-effectiveness of practise.

The process of gathering the evidence to inform practise may progress through several discrete stages. As shown in Table 7.1, Everitt and Wessely (2008) outline five steps towards achieving evidence-based practise.

If the sequence in Table 7.1 is set against the extant literature, what may be concluded with regard to the therapeutic use of gardens and gardening?

TABLE 7.1 Steps Towards Evidence-based Practise (after Everitt & Wessely)

1. **Theory:** How does the intervention relate to theoretical underpinning? What testable hypotheses emerge from theory?
2. **Modelling:** Formulating the details of the intervention and its effects, which may involve small-scale studies.
3. **Exploratory Trial:** Preparation for a definitive trial by identifying appropriate variables and outcome measures, identifying control groups and potential statistical procedures.
4. **Definitive Randomised Controlled Trial (RCT):** A comprehensive randomised control trial that addresses the theoretically relevant hypotheses to strict, replicable standards.
5. **Long-term Implementation:** Do the findings hold true over the long-term when applied in the field?

Theory

A perusal of the literature given to the effects of nature on human functioning reveals that alongside theories from mainstream physiology, psychology and sociology, purpose-built theories have been formulated such as the biophilia hypothesis, the Transactional Model of Stress and Coping and the Theory of Ecological Perception. Yet further, Vella-Brodrick and Gilowska (2022) nominated enhanced wellbeing, cognitive restoration and stress reduction as mechanisms underlying the benefits of nature. Hawkins et al. (2013) nominate two widely used psychological theories that suggest explanations for the stress-reducing and health-improving effect of gardening; these are Attention Restoration Theory (ART) and Stress Recovery Theory (SRT).

Attention Restoration Theory

In formulating Attention Restoration Theory (ART), Kaplan (Kaplan, 1995; Kaplan & Berman, 2010) draws on the distinction between *involuntary* attention and *voluntary* or *directed* attention. Involuntary attention is invoked automatically when something interesting or novel acts, as it were, to attract our attention. Voluntary attention is not automatic and occurs when we consciously make the effort to attend to something that may not be particularly interesting. Given the effort involved, directed attention may bring about cognitive overload and mental fatigue, sometimes with feelings of irritability. With rest we can recover our powers of directed attention and reduce irritability; one way to aid recovery is to switch to involuntary attention within a restorative environment. A restorative environment, Kaplan suggests, holds four main elements that appeal to our involuntary attention. With regard to *fascination,* Kaplan notes, "Many of the fascinations afforded by the natural setting qualify as 'soft' fascinations: clouds, sunsets, snow patterns, the motion of the leaves in the breeze" (p. 174). A restorative natural environment also gives a sense of *"being away"* from the daily grind so refreshing mind, body and spirit. The *extent* of the natural setting, the number of stimuli within it, adds to its restorative powers. This does not mean the setting has to be extensive in size, Kaplan cites the example of the Japanese garden as a miniaturised environment packed with detail. Finally, *compatibility* between the individual and the type of environment is evident in recreational pursuits as diverse as hiking, bird-watching, fishing and, of course, gardening. The empirical research Kaplan cites in support of his theory is furthered by studies showing, for example, the benefits of physical activity in natural environments (Wicks et al., 2022). However, not all evidence is supportive of ART. Trammell and Aguilar (2021) found little effect of natural surroundings on affect or cognition.

Stress Recovery Theory

When we feel threatened by something in our environment we respond with increases in physiological and psychological arousal. When our levels of arousal become uncomfortable, often experienced as anxiety or anger, we call the situation stressful. Experimental evidence presented by Ulrich (1981) and Ulrich et al. (1991) shows beneficial changes in psychophysiological state according to the type of environment. When compared to an urban setting, experience of nature, particularly when it contains water, reduces levels of stress. The basic tenet of Stress Recovery Theory (SRT), that exposure to nature brings about beneficial physiological changes, has empirical support (Grahn, Ottosson, & Uvnäs-Moberg, 2021; Van Den Berg & Custers, 2011; Yao, Zhang, & Gong, 2021). However, Yao et al. make the point that there are shortcomings, such as a risk of bias, with the research evidence. They recommended that future research address these failings to improve quality and form a strong evidence base on which to proceed.

Modelling

Soga et al. (2017) make the point that while their meta-analysis gave a consistent positive result, there remains the task of untangling the causal relationships between gardening and improved health. As discussed above, ART proposes that contact with nature restores cognitive capability via several interactive pathways. It could be added that gardening encourages physical exercise, a healthy diet and social contact, thereby improving physical and psychological health (Zick et al., 2013). It is likely that there are multiple pathways by which gardening may have a beneficial effect. From a review of the empirical literature, Kuo (2015) identified twenty-one potential causal pathways from nature to health involving environmental factors,

TABLE 7.2 Potential Pathways Between Nature and Wellbeing (from Iqbal & Mansell, 2021)

1. Enjoying the different sensory input
2. Calm nature facilitates a calm mood
3. Enhances decision making and forming action plans
4. Enhancing efficiency and productivity
5. Alleviating pressure from society's expectations regarding education
6. Formation of community relations
7. Nature puts things into perspective
8. Liking the contrast from the urban environment
9. Feeling freedom
10. Coping mechanism
11. Anxious if prevented or restricted

physiological and psychological states and behaviour. In reflecting on the state of the evidence, Kuo suggests:

> Deep relaxation, attention restoration and impulse control, sleep, and social ties seem particularly worthy of attention. No doubt some of the plausible pathways identified here will prove either not to contribute substantially to nature's impact on health, or to contribute only under certain limited circumstances. (p. 6)

An Australian study by Iqbal and Mansell (2021) also suggested multiple pathways between engagement with nature and personal wellbeing. From semi-structured interviews with seven female students, a total of eleven themes, shown in Table 7.2, were nominated by the students regarding nature engagement and their wellbeing and mental health.

In order to test the reliability and validity of these suggestions, further empirical research is required. Once articulation of theory and modelling as to how the theory may work is complete, the next step is to undertake the exploratory groundwork for an empirical trial.

Exploratory Trials

An exploratory trial involves nominating independent and dependent variables, identifying outcome measures, finding control groups and reviewing potential research designs and statistical procedures. In producing empirical evidence, a range of research designs of varying degrees of reliability are available. In applied research, the choice of design may be influenced by practical constraints. Hollin (2008) discusses which alternative research designs and methods of data analysis may be confidently relied upon when practical constraints do not allow the use of the strongest research designs.

TABLE 7.3 Methodological Steps in Hypothesis Testing

Step 1. Formulate the hypothesis as a *null hypothesis*, i.e., no effect, and as an *alternate hypothesis*, i.e., effect as predicted.
Step 2. Decide on experimental design and collect appropriate data to test the hypothesis.
Step 3. Set probability level and conduct the proper statistical test.
Step 4. Is the null hypothesis supported or rejected?

In a disparate, uncoordinated fashion the individual studies on the therapeutic effects of nature discussed above have gone some way towards identifying what is important. However, from a research perspective these studies are not all of equal merit. The stronger studies use a range of methodologies including *hypothesis testing*, popular in psychological research. A hypothesis is an assumption, based on theory or modelling, which may or may not be true. For example, the *biophilia hypothesis* noted in Chapter 2 holds that the child has an intrinsic need for nature. Appealing as this hypothesis may be, to be used as the basis for policy, practise and concomitant financial expenditure, it should be supported by robust testing.

The steps in experimental hypothesis testing are shown in Table 7.3. The hypothesis must be stated precisely in terms of what effect the intervention is expected to have. For example, staying with the biophilia hypothesis, we may hypothesise that children who have contact with nature are happier, are better adjusted to life and have a greater respect for nature.

In a typical hypothesis testing study, data are gathered from two or more samples drawn from an appropriate population. In the biophilia example, children with and without contact with nature could be compared on measures of, say, happiness and social development, allowing the hypothesis to be tested by statistical comparison of the data produced by the two samples. The findings from the data analysis indicate whether the difference between the samples is statistically *significant* or *non-significant* when set against the stated *probability level*. A probability level is the degree of chance the experimenter takes in deciding if their statistical analysis allows them to accept or reject the null hypothesis. The traditional significance level in psychology is 0.05 (expressed as $p < .05$), which means taking a 5% chance of making an error in accepting or rejecting the hypothesis. If an error occurs it may be due to incorrectly rejecting the *null* hypothesis, which is a *Type I error;* this is a *false positive*, i.e., accepting a finding as true when it is not. A *Type II error* occurs when the null hypothesis is not rejected when the *alternate* hypothesis is true; this is a *false negative*, i.e., taking a finding as untrue when it is true. Replication studies can increase confidence in the findings from a single study; as the agreement across studies increases, confidence also increases in the robustness of the finding.

Definitive RCTs

In an RCT, participants are randomly allocated to the treatment or comparison group. The purpose of randomisation is to eliminate the risk of *confounding*, i.e., introducing a systematic difference between the groups that may influence the findings. To give an obvious example, if one group is all under 10 years of age and the comparison group is all over 50 years, then age becomes a confound. Other procedures can be introduced into the design. For example, with *single blinding* a participant does not know if they are allocated to the treatment or control group; with *double blinding* both the researcher and participant do not know to which condition an individual participant is allocated. While steps can be taken to increase the practicality of their findings (March et al., 2005), the fact remains that RCTs are expensive and time-consuming to conduct. Nonetheless, it is held that RCTs are the gold standard in research design. Bhide, Shah and Acharya (2018) state the case: "Evidence from randomized controlled trials (RCTs) is considered to be at the top of the evidence pyramid. It is recommended that clinical practice decisions are based on evidence emanating from well-conducted RCTs when available" (p. 380). Although some commentators are wedded to a particular evaluative methodology (e.g., Parker, Bush & Harris, 2014), others are prepared to take a wider view. Ravallion (2020) argues that as no single approach is perfect, so an over-reliance on a particular methodology, including randomised and unrandomised experiments, introduces problems into the evidence base. There is a strong case for using a broad base of research methods to evaluate an intervention, both experimental (randomised) and quasi-experimental (non-randomised) as well as qualitative methods (Diener et al., 2022).

Long-Term Implementation

There are several official bodies that approve and regulate new medical treatments for public use in the United Kingdom; these include the National Institute for Health and Care Excellence (NICE) and the Care Quality Commission (CQC). These bodies are concerned with the effectiveness, safety and cost-effectiveness of a given treatment. Similarly, practitioners are regulated by professional bodies such as the Health and Care Professions Council (HPCP) and the General Medical Council (GMC). It is not clear who would take responsibility for regulating nature-based therapies, although practitioners would be bound by their professional body (assuming they were so qualified).

Research Bias

In Chapter 4 the thorny topic of research bias was raised, noting the comment of Williams and Dixon (2013) that those working and researching in particular fields may be passionate advocates of a particular approach. This is as true of the therapeutic use of gardens as it is of any other field. The literature is littered with advocates for the efficacy of therapeutic gardens in improving a range of behaviours, cognitions and feelings for people of varying age and condition. However, these positive views are over-shadowed by comments about the shortage of and need for more high-quality empirical evidence. Thus, for example, Wang and MacMillan (2013) point to several methodological limitations: the failure to use standardised measures and outcomes makes it problematic to compare across studies; the need for greater specificity in sampling and a need for an increase in methodological and statistical sophistication. In addition, Wang and MacMillan made the comment that "Although scientifically sound research begins with a theoretical or conceptual framework, more than half of the studies ($n = 14$) did not discuss theory to conceptualize how gardening may affect older adults" (p. 173).

There is a clear need for a concerted research programme to answer the question of the therapeutic effectiveness not just of gardens and gardening but also of contact with nature in its many forms. Once this task is completed, it may be possible to identify the active components and therefore design ever-more effective interventions. If these goals can be achieved, then the rewards will follow in terms of less use of pharmacological interventions, economic savings and, most importantly, improved rates of patient recovery.

References

Bhide, A., Shah, P. S., & Acharya, G. A. (2018). A simplified guide to randomized controlled trials. *Acta Obstetrica et Gynecologica Scandinavica*, 97(4), 380–387.

Diener, E., Northcott, R., Zyphur, M., & West, S. (2022). Beyond experiments. *Perspectives on Psychological Science*, 17(4), 1101–1119.

Ernst, E. (2002). A systematic review of systematic reviews of homeopathy. *British Journal of Clinical Pharmacology*, 54(6), 577–582.

Everitt, B. S. & Wessely, S. (2008). *Clinical trials in psychiatry*. Oxford: Oxford University Press.

Grahn, P., Ottosson, J., & Uvnäs-Moberg, K. (2021). The oxytocinergic system as a mediator of anti-stress and instorative effects induced by nature: The calm and connection theory. *Frontiers in Psychology*, 12, 617814.

Hawkins, J. L., Mercer, J., Thirlaway, K. J., & Clayton, D. A. (2013). "Doing" gardening and "being" at the allotment site: Exploring the benefits of allotment gardening for stress reduction and healthy aging. *Ecopsychology*, 5(2), 110–125.

Hollin, C. R. (2008). Evaluating offending behaviour programmes: Does only randomisation glister? *Criminology & Criminal Justice*, 8(1), 89–106.

Iqbal, A., & Mansell, W. (2021). A thematic analysis of multiple pathways between nature engagement activities and well-being. *Frontiers in Psychology*, 12, 580992.

Johnson, W. T. (1999). Horticultural therapy: A bibliographic essay for today's health care practitioner. *Alternative Health Practitioner*, 5(3), 225–232.

Kaplan, S. (1995). The restorative benefits of nature: Toward an integrative framework. *Journal of Environmental Psychology, 15*(3), 169–182.

Kaplan, S., & Berman, M. G. (2010). Directed attention as a common resource for executive functioning and self-regulation. *Perspectives on Psychological Science, 5*(1), 43–57.

Kuo, M. (2015). How might contact with nature promote human health? Promising mechanisms and a possible central pathway. *Frontiers in Psychology, 6*, 1093.

March, J. S., Silva, S. G., Compton, S., Shapiro, M., Califf, R., & Krishnan, R. (2005). The case for practical clinical trials in psychiatry. *American Journal of Psychiatry, 162*(5), 836–846.

Masterton, W., Carver, H., Parkes, T., & Park, K. (2020). Greenspace interventions for mental health in clinical and non-clinical populations: What works, for whom, and in what circumstances? *Health & Place, 64*, 102338.

Parker, R., Bush, J., & Harris, D. (2014). Important methodological issues in evaluating community-based interventions. *Evaluation Review, 38*(4), 295–308.

Ravallion, M. (2020). *Should the randomistas (continue to) rule?* Working Paper 27554. Cambridge, MA: National Bureau of Economic Research.

Soga, M., Gaston, K. J., & Yamaura, Y. (2017). Gardening is beneficial for health: A meta-analysis. *Preventive Medicine Reports, 5*, 92–99.

Trammell, J. P., & Aguilar, S. C. (2021). Natural is not always better: The varied effects of a natural environment and exercise on affect and cognition. *Frontiers in Psychology, 11*, 575245.

Ulrich, R. S. (1981). Natural versus urban scenes: Some psychophysiological effects. *Environment and Behavior, 13*(5), 523–556.

Ulrich, R. S., Simons, R. F., Losito, B. D., Fiorito, E., Miles, M. A., & Zelson, M. (1991). Stress recovery during exposure to natural and urban environments. *Journal of Environmental Psychology, 11*(3), 201–230.

Van Den Berg, A. E., & Custers, M. H. (2011). Gardening promotes neuroendocrine and affective restoration from stress. *Journal of Health Psychology, 16*(1), 3–11.

Vella-Brodrick, D. A., & Gilowska, K. (2022). Effects of nature (greenspace) on cognitive functioning in school children and adolescents: A systematic review. *Educational Psychology Review, 34*(3), 1217–1254.

Wang, D., & MacMillan, T. (2013). The benefits of gardening for older adults: A systematic review of the literature. *Activities, Adaptation & Aging, 37*(2), 153–181.

Wicks, C., Barton, J., Orbell, S., & Andrews, L. (2022). Psychological benefits of outdoor physical activity in natural versus urban environments: A systematic review and meta-analysis of experimental studies. *Applied Psychology: Health and Well-Being, 14*(3), 1037–1061.

Williams, D. R., & Dixon, P. S. (2013). Impact of garden-based learning on academic outcomes in schools: Synthesis of research between 1990 and 2010. *Review of Educational Research, 83*(2), 211–235.

Yao, W., Zhang, X., & Gong, Q. (2021). The effect of exposure to the natural environment on stress reduction: A meta-analysis. *Urban Forestry & Urban Greening, 57*, 126932.

Zick, C. D., Smith, K. R., Kowaleski-Jones, L., Uno, C., & Merrill, B. J. (2013). Harvesting more than vegetables: The potential weight control benefits of community gardening. *American Journal of Public Health, 103*(6), 1110–1115.

CONCLUDING THOUGHTS

Gardens and those who care for them have been present across the globe for millennia. In the main, gardens were designed and constructed on a grand scale and at great cost as the domain of the privileged few, giving pleasure, providing food and employment to the common man. It is only relatively recently that gardens have become part of the lives of the wider population, precipitating a raft of sociological changes. Of course, nothing stands still, so what may we look forward to in the near future? As Cooper (2006) points out, the very act of gardening can be perceived as taking a dominant stance over our environment to create a "deceptive version of nature" (p. 99). Cooper suggests that the perception view of gardening as deception relies on an (unsubstantiated) idealised view of what "nature" should be. Nonetheless, climate change is changing nature and gardeners must adapt to these changes. Thus, we see some of the familiar traditional plants suited to our temperate climate struggling to cope with changes in patterns of rainfall and high temperatures. To respond to climate change, gardeners need to consider which plant species are resilient to more extreme conditions and which, indeed, may help to counter the extremes of a changing world. Thus, there is an increasing availability of "exotic" plants, such as bananas, cyads, some ferns, grasses and olive trees. The challenge to the gardener is to integrate these newcomers into their garden while maintaining its aesthetic appeal and adapting its structure.

It is not just flora that are affected by a changing climate. The wildlife that both live in and pass through our gardens are under threat. There are some animals, such as bees and butterflies, that the gardener prizes for their work pollinating plants; others are include ladybirds, which devour plant-infesting aphids, and frogs and hedgehogs, which help control the snail and slug population.

DOI: 10.4324/9781003289661-9

There are several ways to encourage these helpful animals to take up garden residence: there are "bee hotels" to tempt pollinators into the garden; ladybird houses; and homes designed for frogs and for hedgehogs. A plentiful supply of food and water adds to the habitat's attractiveness.

The garden presents the individual gardener with a range of physical, psychological and social rewards, the latter exemplified by allotments and community gardens, which provide company alongside fresh food. Communal gardens are also to the fore in institutions such as schools, prisons and hospitals. Important as they are, it would be wrong to over-play their generalised and long-term effects. Cairns (2018) warns of a "magic carrot" by which "children's garden encounters will magically reshape society" (p. 517).

While gardens undoubtedly have beneficial short-term effects on people's demeanour, in the absence of robust evidence, it goes beyond the pale to infer wider, longer-lasting benefits to health. This same point applies to the emerging field of green therapies: no matter how eloquently the case may be presented (e.g., Stuart-Smith, 2020), to satisfy therapeutic standards for dissemination of a particular treatment a prescribed standard of evidence must be met. If the promise of green therapies is to be fulfilled, there is a pressing need for comprehensive clinical trials.

To sum up, it can be confidently stated that the garden offers the gardener a connection with time and place: a connection that shifts and changes, giving rise to the common refrain "Ah, wait until you see it next year!"

References

Cairns, K. (2018). Beyond magic carrots: Garden pedagogies and the rhetoric of effects. *Harvard Educational Review*, 88(4), 516–537.
Cooper, D. E. (2006). *A philosophy of gardens*. Oxford: Oxford University Press.
Stuart-Smith, S. (2020). *The well gardened mind: Rediscovering nature in the modern world*. London: William Collins.

RECOMMENDED READING

There were several books that I found to be particularly helpful during the writing, if you are not familiar with them and want to learn more about gardens, here they are.

Cooper, D. E. (2006). *A philosophy of gardens*. Oxford: Oxford University Press.
Edwards, A. (2018). *The story of the English garden*. London: National Trust Books.
Musgrave, T. (2020). *The garden: Elements and styles*. London: Phaidon Press.
Ross, S. (1998). *What gardens mean*. Chicago, IL: Chicago University Press.
Uglow, J. (2004). *A little history of British gardening*. London: Chatto & Windus.
Way, T. (2017). *Allotments*. Stroud, Gloucestershire: Amberley Publishing.
Way, T. (2020). *Suburban gardens*. Stroud, Gloucestershire: Amberley Publishing.

INDEX

Note: Page numbers in **Bold** refer to tables; and page numbers in *italics* refer to figures